Shipboard Electromagnetics

The Artech House Antenna Library

Antennas by L.V. Blake
Shipboard Antennas by Preston E. Law, Jr.
Shipboard Electromagnetics by Preston E. Law, Jr.
Antenna Design Using Personal Computers by David M. Pozar
Microstrip Antennas by I.J. Bahl and P. Bhartia
Interference Suppression Techniques for Microwave Antennas and Transmitters by Ernest R. Freeman
Principles of Electromagnetic Compatibility by Bernhard E. Keiser
Practical Simulation of Radar Antennas and Radomes by Herbert L. Hirsch and Douglas C. Grove

Shipboard Electromagnetics

Preston E. Law, Jr.

Artech House
Boston·London

Library of Congress Cataloging-in-Publication Data

Law, Preston E.
Shipboard electromagnetics

Bibliography: p.
Includes index.
1. United States. Navy—Electronic installations.
2. Electronics in military engineering. 3. Ships—
electronic equipment. I. Title.
VM480.3.L38 1987 359.8′5′0973 87-19293
ISBN 0-89006-247-1

ARTECH HOUSE, INC.
685 Canton Street
Norwood, MA 02062

International Standard Book Number: 0-89006-247-1
Library of Congress Catalog Card Number: 87-19293

10 9 8 7 6 5 4 3 2 1

This book is dedicated to the Memory of

Benjamin Logan Fowler, Sr.

and

Preston Eugene Law, Sr.

Two magnificent fathers: My wife's and mine

"We Make a Living by What We Get
but We Make a Life *by What We Give"*

–Winston Churchill

Contents

The Author

Preston E. Law, Jr., is currently Head of the Ship Topside Electromagnetics Design Branch of the Naval Sea Systems Command where he is responsible for ensuring Electromagnetic Compatibility (EMC) and control of Electromagnetic Interference (EMI) in Shipboard Combat Systems. His past experience includes employment as a Communications Systems Engineer with the Defense Communications Agency in Thailand; the Voice of America in Washington, D.C.; and the Federal Aviation Agency in Anchorage, Alaska. Mr. Law is an electrical engineering graduate of the University of Maryland and a registered Professional Engineer in the State of Maryland. He is a Senior Member of the Institute of Electrical and Electronics Engineers (IEEE), a member of the American Society of Naval Engineers (ASNE), Armed Forces Communications Electronics Association (AFCEA), Association of Old Crows (Electronic Defense Association), Association of Scientists and Engineers (ASE) of NAVSEA, and Tau Beta Pi and Eta Kappa Nu Engineering Honor Societies. He is author of *Shipboard Antennas* (Artech House, Norwood, MA, 1983; Second Edition, 1986).

Preface

Since the end of the nineteenth century, when the science of radio communications was first embraced for sea service, naval ships have never been without electromagnetics. The adoption and evolution of naval shipboard electromagnetics has resulted in a series of marvels from communications to navigation to radar to weapons control and electronic warfare. Yet each of these marvels has been countered step-by-step along the way by that bane of electromagnetic science, interference.

From the initial experimental ship-to-shore wireless tests of 1899 to this very moment, electromagnetic interference has posed a serious and perplexing problem to reliable, effective naval operations. Moreover, after a century of shipboard electromagnetics, electromagnetic interference has changed over the years only in severity and complexity. Today, with a myriad of highly sophisticated electromagnetic systems on board ships, the problem of electromagnetic interference is intense. It has required the marshaling together of the finest of electromagnetic engineering specialists in an all-out fight to bring the interference under control. Continued ingenuity and diligence is absolutely necessary to ensure that the integrity of shipboard electromagnetics, so essential to modern-day electronic weapons and warfare, is sustained.

Preston E. Law, Jr.
15 May 1987

Acknowledgements

Not just for their personal encouragement but more for their matchless engineering contributions to the United States Navy and our country, I wish to acknowledge my coworkers, my colleagues, and associates. Both in government and in private industry, they are toiling diligently in the highly specialized and peculiarly demanding field of shipboard electromagnetics. As dedicated, conscientious experts, they deserve far more recognition than my mere tip-of-the-hat here. Special thanks are due three who carefully reviewed portions of the manuscript: Russell D. Powell (Grounding and Bonding); Neil T. Baron (Radiation Hazards); and James A. Kudzal (Electromagnetic Pulse)—and to our Division Director, James F. Garrett, for his leadership in prodding us to a goal of excellence in service to the fleet.

Also, in preparing the manuscript for this book, I was unusually blessed to have the professional assistance of a first-rate technical editor, Ms. Jean E. Kennedy, and the outstanding quality of Ms. Jennifer A. Love's splendid typography. (One could say in all truth that Jean's and Jennifer's endurance in the preparation of this manuscript was a labor of love.)

Finally, I thank my wife sincerely for her wonderful patience and cheerful interest in my endeavors despite seemingly being left single and alone for many an evening and weekend over the past year.

Introduction

Electromagnetics, in the special sense that we will be concerned with in this book, is the interaction of electric and magnetic fields associated with *radiated* energy—i.e., energy in the form of radiated electromagnetic fields. As such, the principal topic of interest here will be shipboard devices that emit or sense electromagnetic waves in the radio-frequency spectrum from extremely low frequencies through extremely high-frequency microwave. Other portions of the electromagnetic spectrum such as light waves (e.g., infrared and fiber optics) will not be addressed in depth.

Nor is any rigorous mathematical treatment of electromagnetic theory included herein. Since the days of James Clerk Maxwell's classic treatise, exquisite theoretical works have been published in full measure. A representative sampling is included in the Bibliography.

Emission of shipboard electromagnetic energy may be desired, as in the case of communications, navigation, radar, and weapons control systems through associated transmitting antennas. Emission of electromagnetic energy, however, may be (and, unfortunately, too often is) undesired, as in the instances of unsuppressed intermodulation products, harmonic frequencies, broadband noises, spurious signals, impulse noise bursts, high-level sidelobe energy, parasitic reradiation, multipath reflections, and radiation hazards.

Similarly, the sensing of electromagnetic energy can be categorized into that which is planned (desired reception of communications traffic, radar return pulses, navigation position fixes, and weapons control tracking), and that which is clearly unwanted whenever extraneous electromagnetic emissions are picked up or induced as interference.

Naval ships, being so generously equipped with electronic systems, must contend with all aspects of electromagnetic radiation and reception. Very-high-power emitters must coexist and operate simultaneously with ultrasensitive receptors compatibly in the complex shipboard environment. Interference between these many systems must be eliminated, or suppressed to minimum, in order to have each system perform effectively in support of the ship mission. The effort is not easy.

This book endeavors to discuss each of the chief concerns of shipboard electromagnetics and to show how the problems of compatibility and interference are resolved. It is hoped that in this manner the book will provide much needed understanding and assistance to naval engineers working in ship systems design and operations.

Chapter 1
Historical Background

1-0 PRELUDE

In the 16 May 1986 *Washington Post*, and in the 14 July 1986 issue of the weekly *Navy Times*, it was revealed to the American public that the captain of *HMS Sheffield*, the British destroyer sunk by an Exocet missile during the Battle of the Falklands, had his ship's radar turned off so that he could use radio communications back to England. It was because the radar was interfering with his operational phone communications that the captain ordered it shut down, and it was during the time the radar was off that the Exocet came in undetected.[1,2]

The *Navy Times* reporter in particular stressed the point that the problem was one of electromagnetic interference (EMI) between electronic systems, and that the same or similar problems could occur in a US ship. He quotes Navy sources who acknowledge that EMI is not understood as well as it ought to be, and that the Navy has for the last three or four years "been on an electromagnetic compatibility campaign."

The *Sheffield* incident dramatically underscores the serious impact that EMI can have in disrupting or degrading ship electronic systems, especially in crises. And most certainly there is an urgent effort ongoing to control the harmful effects of EMI. However, the problem is not new, nor are the strenuous efforts to contain it satisfactorily.

There is in the very nature of shipboard electromagnetics an essential need for systems electromagnetic compatibility (EMC) and freedom from EMI. Were it not for this inherent relationship and the continuous struggle it breeds, electromagnetics aboard ship would be as satisfying an engineering discipline as any of those ashore, where the luxury of widely separated facilities mitigates many of the compatibility problems. The shipboard situation unquestionably intensifies the issue. Because of this, each of the successes in ship electromagnetics from the beginning (e.g., radio communications, navigation radar, and weapons control) has introduced attendant electromagnetic incompatibilities and interference problems. Thus, in effect, the history of shipboard electromagnetics can be traced by reviewing the efforts made from the very beginning to stem interference and to establish compatibility.

1-1 NAVAL ADOPTION OF ELECTROMAGNETICS

It all began just prior to the twentieth century with that forerunner of modern radio known as "wireless." The new concept of wireless telegraphy, spurned by many skeptics as of no real value, generated considerable interest in the US Navy. Recognizing that wireless might overcome the limitations of visual communications and navigation at sea, particularly in fog, on starless nights, and in foul weather, the Navy, on 26 October 1899, initiated the first shipboard experimental tests. It was the birth of American naval radio communications and of the phenomenon of EMI. As part of the experiments, the Navy Department had the foresight to investigate the use of two transmitters operating simultaneously and the methods used to overcome interference. L. S. Howeth in his definitive *History of Communications-Electronics in the United States Navy*, writes with good humor:

> The results of the interference tests were perfect. That is, the interference was perfect. From time to time the land station transmitted signals while one ship was receiving from the other, which always resulted in utter confusion with the tape being rendered absolutely unintelligible. Concerning this defect it was reported:
>
> "When signals are being transmitted from one station to another, as between the USS NEW YORK and the Highlands Light, and another vessel comes within signaling distance and attempts communication with the Highlands Light, then the signals from the two ships become confused, and the receiving station on shore is unable to distinguish between them."
>
> This was very disappointing, but since the three installations were operating on about the same frequency the result was inevitable. If the same experiment were to be repeated today with broadband transmitting equipment on approximately the same frequencies the result would be the same. The inability to employ tunable equipment at the time was unfortunate, for impressions developed about the inevitability of interference with (wireless) equipment which persisted for years.[3]

So here we have the Navy's initial use of electromagnetics and its immediate troubles with EMI, a natural relationship still with us nearly a hundred years later. It should be noted, too, that EMI was not the only adversity experienced. The naval inspectors were careful to record other potential problems associated with wireless transmission that served as precursors of the electromagnetic hazards with us today:

a. "The spark from the sending coil, or faulty insulation of the sending wire, would be sufficient to ignite an inflammable mixture of gas or other easily lighted matter."[4]
b. "The shock from the sending coil may be quite severe and dangerous to a person with a weak heart."[4]
c. "The sending apparatus and wire would injuriously affect the compass if placed near it."[4]

The first of these potentialities we now refer to as hazards of electromagnetic radiation to fuel (HERF) and to ordnance (HERO). The second is now known as radiation hazards (RADHAZ) to personnel and RF burn hazards. And the third is another of major significance familiar to us as electromagnetic compatibility, or more commonly, EMC.

It is worthwhile to point out here that, depending on one's view, interference may not be always bad. Less than two years after the Navy's first shipboard tests, interference was cleverly exploited for a deliberate gain. In the 1901 International Yacht Races three wire services were to report on events: wireless experts Lee De Forest for Publishers Press Association, Guglielmo Marconi for Associated Press, and John Pickard for American Wireless Telephone and Telegraph. As might be expected, the competition among the three was strong. Howeth recounts the amusing results:

> During the contest both the Marconi and De Forest mobile stations noticed their shore units signaling frantically with flags asking "What is the matter? Signals confused. Cannot read." De Forest tried to improve his transmissions, and, seeing no more signaling, gained the impression he was getting through satisfactorily. When his tug docked he expected to be overwhelmed with congratulations, feeling he had made a great showing against his competitors. However, the event had produced three losers, (the yacht) SHAMROCK II, Marconi, and De Forest. American Wireless, having no sponsor, had nothing to lose and everything to gain by preventing the reception of their competitors' transmissions:
>
> "There is an account that the true culprit in this fiasco was American Wireless Telephone and Telegraph Co., which, upon failing in its efforts to get the press associations to make use of their apparatus in the 1901 yacht races, set up a very powerful station near the Navesink Highlands. Throughout the races they sent out so powerful a stream of electric disturbances that they produced the results previously noted in the Marconi and De Forest reception. Pickard maintains that (American Wireless) did report these races, saying 'And when I say "reported," I mean reported and not what the Marconi and De Forest people call reporting; namely, manufactured news that had no basis of fact whatever.' He stated that

(American Wireless) used a plain aerial, 20-inch Queens coil, and a tulip interrupter minus all weights, so that spark frequency was quite high. They put as much current in the primary as their interrupter would stand and, in so doing, radiated considerable energy . . . Their receiving station was located at Galilee and used aural reception as did De Forest. That, incidentally, gave them an advantage over Marconi with his coherer and inker. Pickard claimed that on the trip down to the race area a bright idea came to him as to the modus operandi to be employed to prevent Marconi and De Forest from receiving the transmissions. He happened to have a newspaper at hand, in which one page had been folded over in printing, so that a large-type headline was superimposed over the fine print of the text. He noted that the small type was almost unreadable but that the headline was undamaged. This gave birth to his idea. Why not use large type—namely long dashes many seconds in duration to smear the small-type ordinary dots and dashes of the competitors? Pickard proceeded to work up a code, which, he said, 'was simplicity itself.' As an example, one long dash of 10 seconds would mean COLUMBIA was ahead; two such dashes would indicate SHAMROCK was in the lead; three, they were neck and neck. Following the first series would come other long dashes from one to nine, identified in the code as conveying common actions taking place. Thus equipped, they were able to get their signals through and interfere with the others. 'Marconi and De Forest didn't have a ghost of a chance and our clever rewrite men made up a nice long story from our coded simple instructions.' Strange as it may seem, they received instructions from Galilee sometime later to split time with Marconi, an order considered cowardly by Pickard. Contacting the MINDORA, the Associated Press boat, with the Marconi so-called apparatus on board as Pickard put it, a liaison was arranged. In relating this incident, the professor tells of his encounter with the president of the Associated Press, 'When some hundred feet away, none other than Melville Stone came on deck with a megaphone and began to berate us. For fully 10 minutes he cussed us, not repeating one word twice, and would probably be cussing us yet if I had not gone below, gotten an egg, and by a lucky throw applied it to him via his megaphone. Incidentally, he stopped cussing, and at the same time the negotiation stopped.' In relating what he called 'The final incident of the race "reporting,"' Pickard said, 'When the yachts crossed the finish line, we held down the key and then continued to hold it down, by the simple method of putting a weight on it. Thus, radiating waves, far from practically continuous, though continuous in our sense of the word, we sailed for our home port, and the batteries lasted for the entire hour and a quarter that we utilized to send the longest dash ever sent by wireless.' Following the races, Pickard returned home via Navesink, where the lighthousekeeper

> showed him around and said, 'Oh, by the way, we had wireless telegraphy here the other day. The Marconi men were here with a little black box like a stock ticker, and paper came out of it with long black lines running down the middle of it. Every few minutes the operator would pick up this tape, look at a few feet of it, swear unholily, tear the tape off, and jump on it.' Of this Pickard stated, 'This was the best appreciation of efforts that I ever received.' "[5]

The above incident may be the first recorded use of EMI employed for personal advantage. Later, as the world went to war, intentional interference was introduced to jam enemy systems, and a whole new science of using interference as an aid developed in what is now known as electronic warfare (EW).[6]

Returning to our review of undesired interference associated with wireless, by 1902 the sudden proliferation of commercial stations and amateur hobbyists with homebuilt wireless sets began causing such a high degree of interference that naval officials urged the Government to regulate all wireless operations. Specifically citing wireless's chief defect as its vulnerability to interference, and of the opinion that the principal use of wireless would be for seagoing communications for many years to come, the Navy Department proposed itself as managing agency for all Government and private wireless stations on or near the coast, in order to prevent mutual interference. After considerable official debate and interservice jousting, agreement was reached. On 29 July 1904, President Theodore Roosevelt signed an Executive Order designating the Navy as controlling bureau for all Government stations (and *all* stations during war; meantime the Department of Commerce would begin regulating private stations). This act firmly established the Navy as the leading developer of radio electronics in our country during the early years. Before the end of 1904, the Navy had 33 ships and 20 shore stations equipped with wireless.

Despite these regulatory attempts, however, the annoying problem of interference continued to increase among the competing Government, private, and amateur stations. Commercial operators, as an example, made unconcealed attempts to prevent each other from transmitting. And many amateurs enjoyed the fun of interrupting both Government and commercial traffic. The result was more requests for effective Government regulation.

Unhappily in the meanwhile, the Navy was experiencing its own difficulties with wireless interference during operations at sea. In 1906, in an effort to evaluate the strategic use of wireless, the Atlantic Fleet conducted large-scale ocean exercises. Because of the limited ranges of the equipment and the vexing effects of interference, the results were quite disappointing. Such failures caused senior naval officials to have strong reservations about the reliability of wireless communications, seriously retarding its development for naval use.[7]

1-2 EARLY ATTEMPTS TO REDUCE WIRELESS INTERFERENCE

The Navy, quick to recognize that spark-generated wideband-transmitted RF energy would have a high likelihood of causing mutual interference, began immediately to evaluate all available equipment prior to deciding which to purchase in quantity. Particular emphasis was placed on good selectivity characteristics in the equipment to reduce mutual interference. To facilitate this task, the Radio Division of the Navy's Bureau of Equipment was established in 1903 and charged with the responsibility of developing and procuring reliable wireless sets that would operate without interference.

By 1906, separately tuned primary and secondary coupling circuits and improved spark-quenching schemes were used to limit the radiation of undesired wideband energy. At the same time, better operating discipline and careful assignment of wireless frequency channels were incorporated.[8]

In 1912 the new name "radio" displaced the older word "wireless." Two years later, during the 1914 occupation of Veracruz, the Navy experienced its first use of radio communications under war conditions. The results were not entirely satisfactory. The spark transmitters of nearby foreign warships generated such heavy interference as to disrupt communications totally. A time-sharing plan had to be worked out among the participants, resulting in US operators being allotted a two-hour period for radio transmission and the other four nations one hour apiece. Thus there was a four-hour interval each day when it was not possible for military headquarters in Washington to be in contact with its forces in the field. Note that an operational method (time-sharing) had to be implemented to avoid the consequences of interference—a method proposed (along with frequency management) as early as 1911,[9] and though better refined in various ways, still used to this day.

The necessity and value of radio was proven many times over, at sea and on land, during World War I. Such widespread usage of course spurred development of equipment improvements. In 1918, superheterodyne techniques in receiver circuitry were introduced to allow broader RF amplification, better selectivity, and much easier operation. It was soon apparent to the Navy, however, that the superhet receiver was far more susceptible to interference aboard ship where many transmitters and receivers were operating in close proximity. It took careful application of shielding technology, circuit isolation, oscillator stabilization, and RF preselection to adopt the superhet satisfactorily to shipboard service.[10]

. It was during this period, too, that means were sought to eliminate the considerable harmonic interference created by the Navy's arc type of transmitter. High-pass filters were added as an arc shunt, and coupling of the arc to a rejector circuit was employed to suppress unwanted emissions. Nevertheless, fleet exercises of 1922–23 clearly showed that arc and spark transmitters generated so much interference that simultaneous shipboard reception was virtually impossible. With requirements for the number of communications channels rapidly increasing, a solution to the interference problem had to be found.

Fortunately, the answer came in the early 1920s with the application of electronic vacuum tubes. Transmitters using tube circuits produced far less RF "trash" than the arc and spark predecessors. At about the same time an anti-keyclick device was adopted to eliminate transient clicks being received during transmitter keying, and thereby affording much closer frequency channel spacing.

1-3 FROM RFI TO EMI

By the 1930s there arose an engineering art devoted solely to the study, measurement, and resolution of radio interference. Naval laboratories and private industry alike sought methods to cope with both the production and the reception of shipboard noise interference. The term radio frequency interference, or RFI, began to be used for describing the nature of undesired electromagnetic emission phenomena, whether by radiation, induction, or conduction. Results of Navy testing, and findings, began to appear in documented journal articles and reports. A representative sampling from the 1930s includes:[11]

a. Bulletin of Engineering Information No. 101, October 1936: "Transmitter Improvement Interference Elimination"
b. Bulletin No. 103, April 1937: "Spurious Interference Responses in Superhets"
c. Bulletin No. 104, July 1937: "Interference Surveys"
d. Bulletin No. 105, October 1937: "Noise Elimination"
e. Bulletin No. 106, January 1938: "Interference and Receiver Selectivity"
f. Bulletin No. 109, October 1938: "Survey of Radio Noise on *USS Yorktown*"

1-3.1 World War II Naval Electronics and RFI

Fueled by the urgent needs of World War II, the 1940s witnessed a tremendous surge in new technology and in the number and type of shipboard electronics systems. RFI problems compounded dramatically. Now air search radars, surface search radars, weapon firing radars, radio navigation systems, and electronic countermeasures equipment vied with radio communications in the congested shipboard environment for a share of the crowded electromagnetic spectrum. Because of immediate operational needs, new equipment had been hurriedly installed without regard to compatibility or interference. While ostensibly increasing the ship's mission capabilities, the mutual disturbance resulting from adding so many new electronic systems soon limited equipment effectiveness. Whereas large prewar ships might have held five or six radio transmitters and a half dozen radio communication and navigation receivers, the large World War II combatant ships carried as many as 12 radio transmitters, 18 receivers, plus a couple of radars. So many high-power radiating and sensitive receiving devices having to operate in proximity created severe intra- and intership RFI.[12] Now the Navy perceived that it had to face a real struggle, and, on 14 June 1945, issued the first joint Army-Navy RFI standard, JAN-I-225, titled *Radio Interference Measurement*.

Analyses of what RFI is and what causes it were promoted, as well as continued improvements to prevent it. Better shielding methods were applied to isolate and contain interference within the source, while at the same time to exclude its entry into susceptible circuits. Filtering networks were devised to reroute RFI away from causing harm.

1-3.2 Postwar Efforts

Official concern about the growing complexity of RFI and the potential catastrophic failures it might produce was demonstrated by the tri-service contract awarded to the Armour Research Foundation (now Illinois Institute of Technology Research) in 1953 to determine the magnitude of the RFI problem and to recommend means for reducing it. Shortly thereafter the First Tri-Service Conference on Radio Frequency Interference, sponsored by government and industry, was held in 1954.[13] Three years later, on 10 October 1957, the Institute of Radio Engineers (now Institute of Electrical and Electronic Engineers, IEEE) established a Radio Frequency Interference professional society. In June 1958, the Navy published its *Electronic Interference Control Manual for Forces Afloat* [11], a handbook to assist in practical application of interference detection and reduction in ships. The manual was a classic for its time as a comprehensive source of shipboard RFI description, causes, and remedies. Further, Appendix 1 of this manual is a valuable bibliography listing naval reports, articles, and field changes from 1936 to 1956 relative to radio interference.

Interest in controlling RFI was gaining rapidly as evidenced by a quote from the 1 October 1958 Fourth Tri-Service Conference on RFI:[14]

> Unfortunately there has been a tendency on the part of many of us in electronics to treat the problem of interference either as a necessary evil or one which would go away if we ignored it. We poured huge human and financial resources into the development of truly marvelous electronic equipments and systems, only to have them rendered, in many instances, completely ineffective because we have failed to apply what would have been a ridiculously small portion of the overall effort to the problem of interference reduction. Today, the folly of this oversight is clearly evident.

Interest in undesired electromagnetic radiation characteristics was not limited to equipment interference effects, however. There was growing anxiety about safety hazards involved with electromagnetic radiation. To review these concerns and set a course of action, the Department of Defense Electromagnetic Radiation Hazards Working Group conducted its inaugural meeting on 30 September 1958. As an outcome of the meeting, the Navy's Bureau of Ordnance was assigned responsibility for developing standards of HERO; the Bureau of Aeronautics was charged with establishing standards for HERF; and the Bureau of Ships was assigned to develop RF RADHAZ technology as follows:

a. Terminology
b. Units of measure
c. Field intensity measurement techniques
d. Instrumentation for measurements
e. Bibliography of papers published in RF RADHAZ
f. Directory of current RF RADHAZ projects.[15]

Just three months later, on 9 January 1959, the Navy conducted its first of a series of power density tests on high-power radiating equipment at a manufacturer's facility. The measurements, done specifically to determine RADHAZ safety zones, were performed on a long-range shipboard UHF air search radar operating at two megawatts peak power with a pulse repetition rate of 300 Hz and pulse width of six microseconds. The test results concluded that the 10 milliwatts per square centimeter safe exposure limit for personnel would be exceeded within 120 feet of the main beam.[16] These initial controlled facility tests served as the foundation for building an extensive library of equipment radiating hazard levels in terms of power density and distance.

Thus, the period following the accelerated electronics growth of World War II was one of examining the many aspects of RFI and of seeking effective ways to contend with it. Admiral Joseph E. Rice, noting the Navy's sponsorship of numerous studies, development of new test equipment, and experimental work in grounding, shielding and bonding, remarked in his opening address to the Tenth Tri-Service Conference on EMC that the 1950s could be characterized as a time of "learning the phenomena" of RFI.[17]

1-3.3 EMC and the Vietnam War Period

The 1960s ushered in a broadening scope and heightened awareness of electromagnetic systems interference. Electronics equipment on typical aircraft carriers, for example, had increased threefold to 35 radio transmitters, 56 radio receivers, 5 radars, 7 navigational-aid systems, and well over 100 antennas.[18] The formal use of the term *electromagnetic compatibility* began appearing, when, in January 1960, the Navy's Bureau of Ships distributed its first *Compatibility of Shipboard Electronics Systems Manual*, outlining procedures for the measurement of radiated RF energy.[19] Of even more significance, in 1961 the Department of Defense established the Electromagnetic Compatibility Analysis Center (ECAC), located at the Naval Engineering Experiment Station in Annapolis, Maryland. The Center was made responsible for applying the newest computer math modeling and data processing analyses for evaluating the electromagnetic environment, developing procedures to increase system compatibility, and reducing the causes of interference and susceptibility.[20]

By the middle of the 1960s, the older term "RFI," which had been in use for 30 years or so, was gradually displaced by the more comprehensive and descriptive expression "EMI." Defined as any undesired radiated or conducted perturbation which degrades the proper operation of electrical or electronic equipment, EMI encompasses a broader spectrum of interest than RFI. As such, the causes of EMI were also becoming much more specific: in addition to atmospheric noises and man-made noises generated by electrical machinery, ignition systems, fluorescent lighting, welding equipment, and circuit breakers and switches, interference resulted from (1) intermodulation noise from mixing of signals in nonlinear transmitter or receiver circuits to create new sum and difference frequencies; (2) intermodulation noise caused by mixing of signals in external nonlinear metallic junctions in the ship structure, rigging, and appendages—principally in corroded or oxidized fastenings and joints, i.e., the so-called "rusty bolt" effect; (3) harmonic and spurious noise products generated in transmitter circuits and not properly filtered out or attenuated; and (4) cochannel and adjacent channel interference present when portions of a signal from one channel penetrate into another.[21] To control the levels of EMI emissions and susceptibility in the design and production of electronic systems, the Navy issued MIL-STD-469, "Radar Engineering Design Requirements for Electromagnetic Compatibility," in 1966, and MIL-STD-461, "Electromagnetic Characteristics Requirements for Equipment," in 1967.

Along with more detailed knowledge of what EMI is came better ideas of how to prevent or suppress it. Methodologies were proposed in the design process for optimizing the topside arrangement of the many electromagnetic emitters and sensors to enhance isolation (RF decoupling) and to reduce degradation. That is, a better understanding of the shipboard electromagnetic environment was strongly encouraged.[22] Moreover, improvements in ship design and construction techniques were urged to reduce EMI, such as the liberal use of nonmetallic materials for lifelines and vertical ladders.[23]

Problems of EMI were noted with much dismay during naval combat operations of the Vietnam War. Recounting the times, Captain J. S. Oller, Jr., USN, wrote:

> By the late 1960s, the magnitude and number of electromagnetic problems were having appreciable effects on Fleet operations to the point where Fleet capabilities were actually constrained by them. Task Force and Unit Commanders were required to take into account the limitations of their electronics interfaces when ordering actions. In some instances, it was standard practice to shut down certain search radars and communications transmitters when missile alert conditions were set in the Gulf of Tonkin. In other instances, aircraft takeoffs and landings dictated such actions. It was a real-life, very constraining environment in which U.S. combatants were operating.[24]

In recognition of these concerns, the Secretary of Defense, on 5 July 1967, signed a directive to establish an integrated Department of Defense program to ensure electromagnetic compatibility. Following this, the Chief of Naval Operations acknowledged the magnitude and seriousness of EMI problems by creating an office of Tactical Electromagnetic Coordinator on 24 November 1969, and made the following statement in his directive:

> One of the Navy's most urgent problems is the management of the electromagnetic environment of naval task forces. Electromagnetic equipment is essential to every mode of naval warfare. In many instances ship and aircraft systems using electronic devices have been developed with inadequate regard for compatibility with the total electromagnetic environment. Electronic planning has in many cases been in the nature of a reaction to meet specific, independent needs. The urgency of immediate problems has in many cases dictated actions without regard to the more involved consideration of systems integration. This frequently has encouraged random proliferation of electronic programs and has created a multitude of budget items in all appropriation categories. As a result, optimization of the electromagnetic environments for both offense and defense has not been achieved.[25]

These high-level moves clearly signalled the Navy's organized scientific and engineering management approach to fighting the battle of EMI in the 1970s and '80s.

1-4 THE MODERN ERA

1-4.1 Emerging Management Interests

In 1970, at the request of the Chief of Naval Operations, a thorough study of electromagnetic problems being experienced in the fleet was initiated. In February 1973, this investigation, known as the Tactical Electromagnetic Systems Study, or more simply the TESS, produced an impressive eleven-volume report which identified over 600 problems. Unfortunately, only limited distribution of the report was made; consequently, no concerted action was taken to resolve the problem. Even though adequately identified, the known problems persisted, and new ones were being introduced upon acquisition of new systems. Quoting again from Oller on the situation existing at that time:

> There are a complex set of circumstances which militated against improvements. Although directives clearly required electromagnetics consideration in acquisitions and for in-service equipments, in real life this fell through the cracks. Acquisition and Program Managers' attentions were on many other major problems, and electromagnetic interference prevention simply was lost in the shuffle. Improvements were also hobbled by the inadequate specifications, constrained by a lack of funds with which to prevent or correct problems, and, lastly, suffered from an insufficient feed-back on either problems in existing systems or those developing in acquisitions.[26]

Coincidentally, at the very time of the TESS report findings, in February 1973, the Navy also launched its ambitious Shipboard Electromagnetic Compatibility Improvement Program (SEMCIP). SEMCIP was chartered to develop standards by which corrective actions required to suppress EMI could be effectively tested, regulated, documented, and promulgated to the fleet. Moreover, SEMCIP was to give broad application to preventive measures for the reduction of EMI, mainly by accomplishing a threefold task:[27]

a. The design and procurement of US naval ships with electronic systems that would be electromagnetically compatible.
b. The identification and reduction of EMI aboard ships currently in the operating fleet.
c. The provision of training to all personnel involved in the design, procurement, installation, maintenance, and operation of a naval ship and its electronic systems to ensure that these individuals have an understanding of the requirements and procedures for achieving and maintaining shipboard EMC throughout the life of a ship.

Here was the first instance of a highly organized quick-response engineering approach dedicated solely to resolving EMI problems being reported by the fleet.

There was continued pressing, too, in the early 1970s, for the adoption of new ship construction techniques to reduce EMI. Innovative developments in welding and joining processes so as to do away with an excess of bolted and riveted joints were emphasized, as were new sealing compounds, gasketing materials, and bonding methods, all for increased corrosion resistance and decreased likelihood of "rusty-bolt" intermodulations. Along with these technologies were additional calls for the replacement of large metallic topside items such as storage boxes, flag bags, stanchions, jackstaffs, and ladders with nonmetallic glass-reinforced plastics; and judicious separation, routing, and shielding of cables in order to preclude EMI pickup and reradiation.[28, 29]

Likewise, very specific electronic circuitry methods were being employed to reduce equipment performance degradation caused by EMI. For example, the following improvements were cited for shipboard surveillance radars in 1976:[30]

a. Prevention of receiver saturation
b. Reduction of false alarm rate
c. Enhancement of signal-to-interference ratio
d. Discrimination of directional interference; e.g., sidelobe jamming
e. Suppression of stationary (slow-moving) clutter

Figure 1-1 illustrates the EMI suppression methods used to achieve the above-listed improvements.

1-4.2 Establishment of TESSAC

A special Tactical Electromagnetic Systems Study Action Council, or TESSAC, was formed in August 1975 "to examine the TESS report and determine the underlying causes for the many unresolved problems, and, finally, to provide a plan of action for resolution of existing problems and prevention of future problems."[31] By querying naval programs, laboratory personnel, and systems engineering directorates to ascertain whether the TESS-reported problems still existed and what remedies had been applied, the TESSAC found that:

a. The majority of Fleet tactical electromagnetic problems which prompted the TESS effort still existed.
b. Inadequate emphasis was being given to tactical electromagnetic considerations in the development and acquisition of new systems and equipment.
c. Known electromagnetic deficiencies in systems and equipment in service were not being aggressively corrected.
d. Directives were being circumvented.
e. Existing management of the tactical electromagnetic effort was being ignored or manipulated so that overall effectiveness was minimal.

INTERFERENCE SUPPRESSION TECHNIQUES	PERFORMANCE IMPROVEMENTS				
	RECEIVER SATURATION PREVENTION	FALSE ALARM RATE REDUCTION	SIGNAL-TO-INTERFERENCE RATIO ENHANCEMENT	DIRECTIONAL INTERFERENCE DISCRIMINATION	STATIONARY CLUTTER SUPPRESSION
AUTOMATIC GAIN CONTROL (AGC)	1	2			
AUTOMATIC NOISE LEVELING (ANL)	1	2			
BEAM-TO-BEAM CORRELATION (BBC)		1	2		
BURN-THROUGH (BT)			1		
DICKE-FIX(DF)	1	1			
FAST TIME CONSTANT (FTC)	1	1			
FREQUENCY AGILITY (FA)			1		
MANUAL GAIN CONTROL (MGC)	1				
MOVING TARGET INDICATION (MTI)					1
NARROW-BAND LIMITING(NBL)	1				
PULSE COMPRESSION (PC)			1		1
PULSE-TO-PULSE CORRELATION (PPC)		1	2		
SENSITIVITY TIME CONTROL (STC)	1	2			1
SIDELOBE BLANKING (SLB)				1	
SIDELOBE CANCELLATION (SLC)			1	1	
VARIABLE REPETITION FREQUENCY (VPRF)		2	1		
VIDEO INTEGRATION (VI)			1		
WIDE-BAND LIMITING (WBL)	1	2			
WIDE-PULSE BLANKING (WPB)		1			1

Note: Numeral 1 denotes primary purpose of the technique, 2 denotes secondary occasional benefit.

Figure 1-1 Shipboard Surveillance Radar Interference Suppression Techniques (1976)[30]

In March 1976, the TESSAC released its recommendations, with particular emphasis on managing the tactical electromagnetic effort, enforcing existing policy, and ensuring the implementation of existing directives. In recognition of its continued need, the TESSAC was asked to continue its work and was directed to investigate contemporary electromagnetic effects; to determine the capabilities of naval laboratories and engineering commands to correct electromagnetic deficiencies; to determine adequacies of specifications and standards in electromagnetic effects; to develop detailed plans to ensure the consideration of deleterious effects of EMI throughout the acquisition process; and to develop electromagnetic technology research and development programs.

The results of this work were summarized in the TESSAC report of September 1977. The report noted that: (1) the current state of technology was viewed as adequate to prevent or reduce most of the Navy's electromagnetic problems; (2) capabilities varied among analysis, testing, prediction and instrumentation, with the depth of manpower insufficient; and (3) specifications and standards were unanimously viewed as the weakest area of all, being cited as not satisfying the need for electromagnetic controls in acquisition, as overlapping, as contradictory, as noncurrent to technology, and as impractical to implement. As previously, the TESSAC stated its opinion that policy and implementing directives were adequate to provide for necessary inclusion of electromagnetic considerations. The primary recommendation of the Council was that the Navy ensure that policy and directives be complied with, that funding be provided, and that cognizant commands establish and support programs adequate to handle electromagnetic problems effectively on a continuing basis.

1-4.3 Implementation of EMC Management

Reacting to Chief of Naval Operations policy guidance, and likely anticipating the forthcoming recommendations of the TESSAC, an instruction was issued on 13 January 1977 to implement EMC management procedures at the ship systems command level.[32] The instruction directed that an Electromagnetic Compatibility Program Plan (EMCPP) be prepared upon initiating development of all electronics equipment and systems designs which involve electromagnetic radiation. Furthermore, planning, programming, and contractual documentation must provide for EMC requirements, analyses, measurements, test and evaluation, and all applicable standards and specifications must be invoked. An EMC Advisory Board (EMCAB) must be instituted for all ship and major systems programs during the design, acquisition, and construction phases for review, advice, and technical consultation on all electromagnetic aspects to identify and resolve potential electromagnetic problems. All ship alterations (SHIPALTs), equipment field changes, engineering change proposals (ECPs), and requests for waivers must include an EMC impact statement. All new electronic equipment and systems must be subjected to thorough EMC analyses prior to commencement of development to ensure electromagnetic compatibility with the operational environment. Further, EMC training and education must be provided for naval and contractor personnel. Program managers must ensure that adequate funding is requested to perform required EMC analyses and measurements to comply with the requirements and provisions of the instruction, and to resolve existing and anticipated fleet EMC problems.

This instruction made it clear in no uncertain terms that henceforth EMC would never be an afterthought in ship design or equipment development for the Navy. An electromagnetic doctrine for the modern Navy was firmly established from that point.

As an aid to better understanding the causes and effects of EMI, in June 1977 the Navy published the *Commanding Officer's Guide to the Shipboard Electromagnetic Environment*. This milestone document discussed typical examples, and sources of EMI and the preventive and corrective measures taken to minimize EMI degradation. A little over a year later, in September 1978, a second publication followed, entitled *The Electronic Material Officer's Guide to Shipboard Electromagnetic Interference Control*, to provide technical information and management procedures helpful in the performing of EMI control functions.

Also in 1978, as an adjunct to SEMCIP, a new plan of action was introduced at the shipyard level called the Waterfront Corrective Action Program, or WCAP. The successful application of EMI solutions learned through SEMCIP would now be institutionalized in the yards to ensure that surface ships would be repaired, overhauled, and maintained in a manner to improve EMC.[33] Training and awareness material, standardized procedures, data files of known problems, and improvements of specifications would be developed to implement and extend the life span of shipboard EMI control. Typical WCAP technical assistance was offered to include:[34]

a. Selective bonding and grounding—such items as inclined ladders, climber safety rails, lifelines, stanchions, metallic flagstaffs and jackstaffs, expansion joints, tilting antenna mounts, and safety nets.
b. Shielding—such as mast-mounted cables against main beam radiations from radars.
c. Blanking—such as the employment and proper programming of pulse-activated blankers with radar directors and EW receivers.
d. Use of glass-reinforced plastics or other nonmetallic materials as selective replacements for lifelines, ladders, boat spanner wires, preventer stays, boat gripes, and flag jackstaffs.

e. Selective replacement of ferrous hardware topside and in antenna near fields with nonmagnetic materials.
f. Insulating—such portable items as fog nozzles, davits, lifelines, booms and personnel stretchers to prevent metal-to-metal contact.

1-4.4 Rising Interest in EMP

Toward the end of the 1970s, another form of EMI began to raise growing concern for naval shipboard systems—that of electromagnetic pulse, or EMP.[35] Generated by the high-altitude detonation of a nuclear warhead, the extremely high levels of field intensity in EMP could prove catastrophic to the very sensitive microminiature solid-state circuit components employed widely in shipboard equipment. As a consequence, new technologies in shielding and in surge protection devices were being developed and incorporated to harden ship systems against the potential effects of EMP.

1-4.5 The Current Status

By the 1980s the Navy had become well accustomed to the phenomena of shipboard electromagnetic interference. More than eighty years of experience had made EMI both a familiar and an expected challenge. Procedures are now quite well known about how to recognize and measure EMI for what it is, and management methods on how to contend with it are explicitly stated as mandatory policy throughout the Department of the Navy. Foremost, it is currently well established that control of EMI has to begin with the electronic design engineer:[36] Each designer of a component or circuit or new equipment or entire electronic system must be aware of, and use, all available means to control EMI. Then, upon completion of the design, the device must be thoroughly subjected to tests for evidence of EMI generation (or susceptibility). Last, the system integration engineer must consider the electromagnetic environment in which the device must operate, and the installing engineer must conform to exacting methods to minimize EMI. This process is essential to affording the equipment and systems at least an opportunity to operate effectively in performing the intended mission, and it results in much saving of time and money. Making corrections after the fact is costly.

Secondly, there is now a strong emphasis on documented requirements. The operational requirement (OR) for any system should define the electromagnetic environment, friendly or hostile, in which the system will operate. Further, the implementation plan should identify system vulnerability to EMI and means to reduce the risk. The Development Proposal should address methods for obtaining the specified levels of EMI control. The Top Level Requirements should state the amount of acceptable EMI degradation. The Test and Evaluation Master Plan (TEMP) should specify the appropriate testing to ensure that required operational characteristics are met. Similarly, the Request for Proposals must include the anticipated electromagnetic environment, the performance requirements in that environment, and the electromagnetic test, evaluations, analyses, simulations, and data to control EMI.

Finally, the Electromagnetic Compatibility Program Plan is the top-level management document for EMC during the design and acquisition. This Plan is used primarily by the design and procuring activity to ensure that all pertinent EMC considerations are implemented throughout the acquisition program, including the means for EMI control, from start through final design and production and throughout the operational life of the equipment.

In the area of practical applications, the 1980s have seen several new improvements. For one thing, hardware solutions that have proven successful for specific problems and are seen as applicable to commonly experienced troubles have been developed into generic standardized modular units. These add-on interference suppression modules are used to correct shipboard EMI deficiencies. They include such items as time and frequency blankers, notch filters, signal processors, broadband interference cancellers, self-interference cancellers, and a chemical bonding agent to reduce intermodulation by neutralizing nonlinear corroded junctions.[37]

Another important innovation for reducing the effects of EMI is the renewed interest in use of radar absorbing material (RAM). The unique ability of RAM to absorb RF energy makes it particularly useful for the decoupling of closely located electromagnetic systems and for the reduction of reflected (multipath) electromagnetic energy from ship structures. Because of these meritorious features, RAM is becoming an indispensable engineering technique for control of shipboard electromagnetic degradation.[38]

The 1980s have witnessed, too, a remarkable surge in the application of computer modeling as an aid in enhancing shipboard EMC. Color-graphic illustrations are rapidly generated to display prospective performance and degradation as a function of system integration. The designer is able to discern immediately the advantages, or pitfalls, in varying arrangements of electromagnetic systems, and then to present the rationale for recommended options visually.

So many years of naval experience with the causes, effects, and resolutions of shipboard EMI have resulted in the accumulation of an enormous amount of data. To facilitate efficient use of this data, the Navy has implemented a computerized data management program for EMC design feedback and analysis.[39] This automated data base provides a unified system of collecting, consolidating, reporting, and analyzing EMI problems. The data is then stored for feeding information back into the ship system design and procurement process. In this manner it is hoped that the resulting lessons learned will systematically preclude recurrence of the problem.

1-5 CONCLUSION

Nearly a century ago the Navy eagerly became the first user of radiated electromagnetic energy in America. It was a wise decision—remarkably astute, for that original need of wireless communications aboard ships has proven absolutely essential ever since. From our twentieth century perspective, shipboard communication is accepted as an inherent part of naval ship design and operations. Moreover, it appears destined to be so as long as there is a Navy.

Yet, at the very instant of accepting wireless electromagnetics as an operational need, the Navy unwittingly accepted the unwanted phenomenon of EMI. Thus, these two opposing natures, electromagnetics as an asset and electromagnetics as an interference, have evolved together from the simplistic days of wireless radio to the present sophistication of a virtually electronic Navy. Doubtless, the naval scientists and officials who subscribed to wireless for ships at the dawn of electromagnetics would be utterly astonished, if not petrified, to see today what man and nature have conspired to create together.

And just where are we today, after so long and complex an electromagnetics evolution—opposed at every stage of development by insidious modes of interference?

> Navy ships today could not function without electronics. Electronics provide communications, command and control, navigation, radar surveillance and tracking, weapons control, and data processing. With so many systems competing for scarce portions of the finite frequency spectrum as well as for the limited space aboard ship, [there are] serious problems in trying to make the systems work well together.[40]

But work together they must! And as we have seen in this historical overview, the Navy has had to develop an entire doctrinal policy to see that shipboard electromagnetic systems do indeed work together effectively. It has been a long, hard battle, and the tide has turned in our favor:

> The Surface Navy is on the verge of having its electromagnetics act together on Navy ships. For the first time the necessary assets are coming in place:
>
> - Organization
> - Funds
> - Authority—Workable policy is in place at all levels.
> - CNO Support—The Chief of Naval Operations has personally approved EM program progress.
> - Fleet Recognition and Support—The Fleet has taken on training and self-help responsibilities and is actively implementing EM control.
>
> However, the job is not done, nor are we even past the bow wave. Continued active defense and use of these assets is required. The potential is

clearly there to produce and modernize ships which fully utilize their electromagnetic systems and have maximum combat capability.[41]

To assure electromagnetic compatibility among all the sensitive electronic equipment installed on naval platforms will require careful attention to potential EMI problems by the entire shipbuilding community: the designer, the builder, and the operator. The ability to establish workable compatibility is in place and becoming common practice.[42]

REFERENCES

1. "Captain's Phone Call Jammed Ship's Defense in Attack," *The Washington Post*, Washington, DC, 16 May 1986.
2. F. Elliott, "Navy Learns From Brit's Error in Turning Off Radar," *Navy Times*, Springfield, VA, 14 July 1986.
3. L. S. Howeth, *History of Communications-Electronics in the United States Navy*, U.S. Government Printing Office, Washington, DC, 1963, p. 32.
4. *Ibid.*, p. 34.
5. *Ibid.*, pp. 38–39.
6. P. E. Law, Jr., "Electronic Warfare," *Shipboard Antennas*, Artech House, Dedham, MA, 1983, pp. 413–417.
7. Howeth, *op. cit.*, pp. 69–83; 117–124; 521.
8. L. A. Gebhard, *Evolution of Naval Radio-Electronics and Contributions of the Naval Research Laboratory*, Report 8300, Naval Research Laboratory, Washington, DC, 1979, pp. 5–6.
9. D. W. Todd, "The Navy's Coast Signal Service," *Journal of the American Society of Naval Engineers*, Vol. 23, No. 4, 1911, pp. 1101–1102.
10. Gebhard, *op. cit.*, p. 61.
11. "Electronic Interference Control Manual for Forces Afloat," *International Electronics Engineering Inc (IEEI) Report 8031*, June 1958, Appendix 1, p. 9.
12. K. D. Wilson, "Communications, A Limiting Factor In Naval Warfare," *Naval Engineers Journal*, February 1963, p. 52.
13. S. I. Cohn, "Electromagnetic Compatibility," *American Society of Naval Engineers Journal*, November 1961, pp. 763–764.
14. L. W. Thomas, Sr., "EMC Twenty-Five Years Ago," *IEEE Transactions on Electromagnetic Compatibility*, Vol. 25, No. 3, August 1983, pp. 133–134.
15. G. J. Marks, "Minutes of the First Meeting of the Electromagnetic Radiation Hazards DOD Working Group," Department of Defense Armed Forces Supply Support Center, Washington, DC, 16 October 1958.

16. "RF Radiation Hazards Measurements on Long Range Air Search Radar," *Department of the Navy Bureau of Ships Task 2-1-59, IEMC Report 9008*, Washington, DC, 9 January 1959, pp. i, 1–2.
17. ADM J. E. Rice, USN, "Compatibility at the Turning Point," Keynote Address, *Proceedings of the Tenth Tri-Service Conference on EMC*, November 1964.
18. Wilson, *op. cit.*, p. 53.
19. "Instrumentation Techniques for Radiation Hazards Measurements," *Compatibility of Shipboard Electronic Systems*, RCA Report 311-60, Department of the Navy Bureau of Ships, January 1960.
20. Cohn, *op. cit.*, pp. 764–765.
21. H. M. DeJarnette and H. J. DeMattia, "Electromagnetic Interference Considerations for Shipboard Electronic Systems," *Naval Engineers Journal*, June 1966, pp. 459–464.
22. CAPT M. Eckhart, Jr., USN, "Topside Design for Electromagnetic Effectiveness," *Naval Engineers Journal*, June 1969, pp. 57–65.
23. CDR J. V. Jolliff, USN, "Research and Development: The Key To The Electromagnetic Interference Problem," *Naval Engineers Journal*, December 1969, pp. 52–57.
24. CAPT J. S. Oller, Jr., USN (RET), "The Navy's Tactical Electromagnetic Program—Problem or Panacea," *Naval Engineers Journal*, October 1978, p. 51.
25. *Ibid.*, p. 52.
26. *Ibid.*, p. 53.
27. J. C. P. McEachen and H. K. Mills, "The Shipboard Electromagnetic Compatibility Improvement Program (SEMCIP)—A Program for the Operating Fleet," *Naval Engineers Journal*, October 1976, p. 65.
28. "Electromagnetic Influence on Shipbuilding Performance Specifications and Practices," Naval Ship Engineering Center, Hyattsville, MD, IITRI Report E6181-F, December 1970.
29. CDR J. V. Jolliff, USN, "Improvements in Ship Construction Techniques to Reduce Electromagnetic Interference," *Naval Engineers Journal*, April 1974, pp. 13–27.
30. R. J. Biondi, *Handbook for Shipboard Surveillance Radar*, NAVSEA 0967-LP-587-0010, Naval Sea Systems Command, Washington, DC, January 1976, pp. 41–42.
31. Oller, *op. cit.*, pp. 53–54.
32. *NAVSEA Instruction 2410.2*, Department of the Navy, Naval Sea Systems Command, Washington, DC, 13 January 1977, pp. 1–5.
33. D. B. Winter and R. V. Carstensen, "Waterfront Corrective Action Program (WCAP) Pilot Shipyard Project: Resolving EMI Problems in Con-

junction with the Ship Overhaul Process,'' *Naval Engineers Journal*, February 1982, p. 41.

34. LCDR J. N. Layl, USN, and CDR G. D. Ellis, USN (RET), ''A Fleet Oriented Electromagnetic Interference Control Program,'' *Naval Engineers Journal*, February 1982, p. 48.
35. R. V. Carstensen, ''An Approach to EMP Hardening for Naval Ships,'' *Naval Engineers Journal*, April 1979, pp. 141–154.
36. J. F. Garrett, R. L. Hardie, and P. A. Rogers, ''Let's Design Out EMI,'' *Naval Engineers Journal*, February 1982, pp. 37–40.
37. J. F. Garrett and R. J. Haislmaier, ''EMC For Forces Afloat,'' *1983 IEEE International Symposium on Electromagnetic Compatibility*, 23–25 August 1983, pp. 113–117.
38. R. T. Ford, ''Enhancing Electromagnetic Compatibility On Naval Ships With Radar Absorbent Material,'' *1983 IEEE International Symposium on Electromagnetic Compatibility*, 23–25 August 1983, pp. 130–133.
39. J. F. Garrett, A. Jackson, J. Kelleher, and A. Light, ''EMC Design Feedback and Analysis Data Management Program,'' *1983 IEEE International Symposium on Electromagnetic Compatibility*, 23–25 August 1983, pp. 118–122.
40. Garrett and Haislmaier, *op. cit.*, p. 113.
41. J. C. P. McEachen, ''On Getting Our EM Act Together and Putting It on a Navy Ship,'' *Naval Engineers Journal*, August 1982, p. 57.
42. Garrett, Hardie, Rogers, *op. cit.*, p. 37.

Chapter 2
The Shipboard Electromagnetic Environment (EME)

2-0 THE TANGIBLE ENVIRONMENT

The topsides of modern naval surface ships have been aptly described as environments of multiple electromagnetic scattering obstacles. To anyone intimately familiar with the concept of EMC, who has spent any length of time above deck on a Navy ship, that description is visually definitive. There have been other, less elegant, illustrations offered, ranging from ''electromagnetic jungle'' to ''electromagnetic nightmare.'' Certainly all would agree that it is a most unfriendly environment for the well-being and good operation of electronic systems.

First, there is simply all that passive metal. A host of inert metallic projections greets the eye: exterior bulkheads, inclined ladders, stanchions and booms, mast legs and yardarms, chocks and bits, stacks, cranes, boat davits, storage racks and lockers, handrails and lifelines, flag staffs, cable rigging, upright hatch covers, gun mounts, weapons launchers, and, of course, a multitude of antennas of every sort. (See Figure 2-1.)

These objects, arrayed in an extraordinary mixture of shapes and sizes, act in every conceivable manner to block, intercept, conduct, reflect, scatter, diffract, and reradiate electromagnetic energy—and sometimes to create new electromagnetic products in the form of intermodulation interference. There is no escape. A single electromagnetic emitter or sensor might be placed at the very top of the highest mast. A couple of others might be stacked vertically a short distance below in an around-the-mast circular fashion. But all others must suffer from the detrimental effects of the mass of passive metallic objects—those multiple electromagnetic scattering obstacles.

Figure 2-1 Topside of Modern Warship (Numbers Indicate Individual Antennas)

Then there is all that electrically active metal; i.e., machinery devices being powered by motors and generators to operate tools, cranes, and booms; to point weapons systems; and to rotate antennas. These electrical entities not only augment the family of metallic obstacles, they also contribute mightily to the onboard ambient electromagnetic interference.

Finally, there is the matter of the natural marine environment. Exposed to the atmospheric elements and to battering seas, the topside of a ship is subjected to near continuous coatings of salt spray. Such moisture, particularly when mixed with stack gas contaminants, promotes early corrosion and rapid physical deterioration.

Thus combined, so much metal, so congested and confined, in so harsh a nature, can result only in a clearly hostile environment for topside electronics systems. We have not yet even mentioned the deleterious effects of invisible contributors—the wildly varying electromagnetic radiating fields adding to the overall environment. No wonder, then, that it takes a corps of highly trained EMC specialists to cope with shipboard electromagnetic design and integration.

2-1 THE COMPOSITE RF ENERGY ENVIRONMENT

For the specialist, there is much more to the shipboard electromagnetic environment than meets the eye. The unseen, too, must be grasped and dealt with in all its many forms. It is the invisible RF medium that makes the problem so much more difficult.

The shipboard RF environment is a complex mixture of radiated electromagnetic energy created from multiple sources. The chief contributors are onboard emitters, comprising: (1) HF communications transmitters, (2) VHF communications transmitters, (3) UHF communications transmitters, (4) satellite communications transmitters, (5) air search radars, (6) surface surveillance radars, (7) surface navigation radars, (8) air control radars, (9) weapons directing radars, (10) electronic warfare jammers, (11) identification, friend or foe (IFF) transponders, and (12) tactical air navigation (TACAN) homing beacons.

Bear in mind that the ship transmitters cited above are all onboard *intentional*, desired radiators of RF energy. Also present in the shipboard environment are intentional incoming RF transmissions (e.g., communications and navigation data) from friendly external sources, and, in most circumstances, many forms of *unintentional* extraneous RF emissions from nearby friendly sources (e.g., ships of the fleet operating in proximity). Add to these the potential for undesired deliberate RF transmissions from unfriendly sources (e.g., enemy surveillance and jamming). Finally we must include the natural RF interferences (lightning, galactic, and atmospheric noise) and man-made interference emanating from electrical machinery and components. The composite total of this transparent RF medium, it can be appreciated, is very complex indeed. Into this environment we immerse sophisticated and sensitive electronic systems, demanding that they perform effectively.

2-2 EFFECTS OF THE SHIPBOARD EME

Because of the nature of the shipboard electromagnetic environment, no major naval ship is completely free of its adverse effects. Some degradation, even if mild, will always be evident. It is the task of EMC design and integration (discussed in the next chapter) to ensure that each electronic system operates effectively despite the degradation experienced within the intended shipboard EME.

It must be stressed, moreover, that severe forms of electromagnetic systems degradation do occur frequently. Therefore, steps must be taken to suppress and control such problems lest the degrading effects result in serious disruptions, performance errors, or system shutdown.

In general there are two principal causes of electromagnetic degradation. The most basic is undesired strains of RF energy received openly through antennas and transmission lines to gain entry into receiving equipment and systems. The second is unintended penetration of EMI into victim equipment via unsuspected ports. The easier of the two problems to correct is the first, by proper design and "hardening" of the receiver entrance circuitry. The second type of problem is, however, likely to be quite difficult to correct, as it usually requires extreme care to detect and suppress.

It would be well to point out here, in simplistic terms, that for EMI to be experienced, there must be: (1) an interference signal-generating source, (2) a coupling path from interference source to victim equipment, and (3) a system that is susceptible to the interfering signal and its degrading effects. Depending on the equipment, susceptibility characteristics such as amplitude, frequency, and response time vary widely. For example, the victim in question may be narrowly frequency selective or it might be a type receptive to broadband unfocused noise. Some victims may have microsecond response time to peak bursts of energy, while others will react slowly to average signal levels and heating. Thus the susceptibility characteristics, along with the selection of components and suppression techniques such as filtering and shielding, must all be carefully considered when analyzing the unfavorable effects of the shipboard EME.

Typical examples of ship system performance degradation resulting from the EME include:

a. *False Targets*—Experienced on radar display scopes due to HF transmissions coupled from antenna to cables and waveguides. Also from multipath microwave reflected energy received by radar antennas.
b. *False Alarms*—Causing sensitive automatic control systems of ship propulsion systems to shut down. Due to HF transmissions coupled into cables to below-deck compartments, and due to EMI generated by below-deck machinery.
c. *False Bearings*—Generated in TACAN beacon navigation information. Caused by energy reflections from nearby mast structures and by HF transmissions via equipment cabling.
d. *False Tuning*—Undesired and erratic tuning of antenna couplers, caused by close-proximity energy coupled from like equipment located nearby.
e. *Distortion of Communications*—High data error rate and noisy audio communication, caused by hull-generated "rusty bolt" intermodulation interference and by antenna-to-cable coupling of HF- to UHF-receiving equipment.
f. *Distortion on Display Scopes*—Spoking and picture eradication on radar screens, caused by antenna-to-antenna coupling of navigation radar energy to air-control radar receivers, and by HF transmission coupled into radar cabling.

g. *Radiation Pattern Blockage*—Experienced chiefly in omnidirectional systems such as HF, VHF, and UHF communication, and in rotation of directional systems such as radars, EW, TACAN, and satellite communication (SATCOM). Caused by multiple obstructions in the radiation field. Results in loss of coverage and range in the direction of blockage.
h. *Radiation Hazards*—Dangerous levels of electromagnetic field exposure to personnel, fuel, and ordnance due to high power concentrations of RF energy in the topside environment.

The cumulative effects of these types of performance and equipment problems have been known to result in serious mission delays and aborted exercises and to gain the immediate attention of headquarters personnel.

2-3 EME CONTROL TECHNIQUES

Proper control of shipboard electromagnetic environmental interference is essential to ensure effective performance of ship electronic systems. The topic is so large and important as to warrant detailed discussion in Chapter 4. Suffice it to say here that good control is first accomplished by: (1) recognizing the problem as interference degradation, (2) identifying the interference source and means of coupling, and (3) taking the necessary action to correct the problem. Over the years shipboard experience in dealing with EMI in its many, often subtle, forms has resulted in generalized methods to mitigate and control it. These include:

a. *Decoupling*—Decreasing the offending energy level by use of physical distance. For example, providing wide separation between high power emitters and broadband sensors.
b. *Frequency Management*—Careful selection and assignment of operating frequencies to avoid mutual use interference among onboard and task force intership electromagnetic systems.
c. *Shielding*—The prevention of interference energy emanations and the reduction of interference susceptibility.
d. *Grounding and Bonding*—The preclusion of conduction of unwanted electromagnetic energy into susceptible equipment, and the neutralization of electrical potential differences between metallic surfaces and joints.
e. *Filtering*—Blocking the passage of undesired energy and passing only desired signals.
f. *Blanking*—Blocking the reception of direct energy radiation by use of electronic pulsed switching circuitry.
g. *Element Arrangement*—Optimum placement of electromagnetic systems in the ship topside to minimize radiation pattern blockage, RF energy reflection and reradiation, and radiation hazards to personnel, fuel, and ordnance.

h. *Antenna Reduction*—Use of multicouplers and multifunction arrays to lessen the number of onboard antennas.
i. *Power Reduction*—Operating at lower emitter power levels to lessen sensitivity degradation of sensor performance.
j. *Metallic Reduction*—Use of nonmetallic materials throughout the ship topside to lessen the number of energy scattering obstacles.
k. *RAM*—Employing RAM to prevent energy multipath reflection and reradiation by absorbing or highly attenuating undesired RF emissions.

Application of these techniques to the shipboard environment for controlling EMI will be examined fully in Chapter 4.

2-4 PREDICTING THE SHIPBOARD EME

It frequently happens that plans are formulated to install newly developed or improved electronic systems into an existing ship environment. Sometimes the installation is an upgraded replacement, and at other times it is an addition. In either event, the electromagnetic characteristics of the new system are fairly well known, along with the ship EME into which the equipment is to be integrated. Thus, an evaluation of the impact of the integration can be made beforehand and verified by actual testing and analysis after the work is accomplished. Accordingly, additions or deletions of equipment, relocation of antennas, or modifications to the ship structure result in the need for continuous updating of the active shipboard EME characteristics.[1]

The more difficult problem, however, is to predict and define a projected shipboard EME; that is, for the case of a totally new ship design and combat systems integration program. We do know that to achieve system compatibility with the environment we must define the EME well into the future so as to cover the entire life span of the proposed equipment.[2] This must include both the equipment and systems parameters and the operational employment, sufficiently described to afford definition of the anticipated EME, as well as the resulting impact of integration of systems into the EME. A threat analysis of the friendly and hostile EME expected to be encountered by the ship must also be performed.

The task is complex and requires the assistance of known and predicted EM data such as that found in MIL-HDBK-235, *Electromagnetic (Radiated) Environment Considerations for Design and Procurement of Electrical and Electronic Equipment, Subsystems, and Systems*, samples of which are depicted in Figures 2-2 and 2-3. Here the EM environment levels are presented in terms of peak and average power density and field strength; it should be noted, however, that there are many other EM-related factors that will influence systems performance. These include antenna characteristics such as aperture, polarization,

LOCATION	FREQ RANGE (MHz)	APPROXIMATE NEAR FIELD EM LEVELS			
		Power Density (mW/cm^2)		Field Strength (V/m)	
		Peak	Avg	Peak	Avg
Table IV - Hangar Deck (CV's and CVN's)	<30 30-2000 >2000	- - -	- - -	32 50 334	10 5 10
Table V(a) - Flight Deck of Aircraft Carriers (CV's and CVN's)	<30 30-2000 >2000	- - -	- - -	200 5100 9700	100 183 183
Table V(b) - Weather Decks, Missile Launching Ships (CG, CGN, DDG, FFG & FF's)	<30 30-2000 >2000	- - -	- - -	200 5100 9700	100 183 183
Table V(c) - Weather Decks, Non-Missile Combat Ships	<30 30-2000 >2000	- - -	- - -	200 5100 7220	100 183 183
Table VII - Envelope of Maximum EM Environment Levels In Main Beam of US Shipboard Emitters	<30 30-2000 >2000	0.11 2000 125,000	0.11 60 410	20 4120 31,000	20 460 300

Figure 2-2 Onboard EME Energy Levels (Approximated)

SOURCE	FREQ RANGE (MHz)	APPROXIMATE NEAR FIELD EM LEVELS			
		Power Density (mW/cm^2)		Field Strength (V/m)	
		Peak	Avg	Peak	Avg
Table I - Maximum EM Environment Levels for Hostile Shipboard Emitters	<30 30-2000 >2000	0.4 14,500 250,000	0.4 90 450	40 7300 30,000	40 600 1400
Table II - Maximum EM Environment Levels for Hostile Airborne Emitters	<30 30-2000 >2000	- 2510 50,000	- 4 65	- 3100 14,000	- 125 500
Table III - Maximum EM Environment Levels for Hostile Landbased Emitters	<30 30-2000 >2000	4 700,000 800,000	4 7000 275,000	120 55,000 850,000	120 5500 33,000
Table X - Actual Hostile Jammers	<2000 >2000	25 35	2 30	300 360	85 320
Table XI - Postulated Hostile Jammers	<2000 >2000	4500 35,000	25 350	4100 12,000	300 1200

Figure 2-3 External EME Energy Levels (Approximated)

pattern form, and scan rate; radiated emission characteristics such as pulse width, repetition frequency, and rise and decay time; arrangement and relative proximity of electromagnetic emitters and sensors; and total EM spectrum being utilized.[3]

Thus, in order to evaluate the overall effects, all known information of the projected EME has to be gathered. Yet the case of new ship design remains quite perplexing because, despite what may be known about equipment and systems characteristics, so much else of the ship's EME is not quantifiable. Even the hull and structure undergo continual changes during the various phases of design, altering a major portion of the passive EME (the ship's topside) at each stage. Moreover, since ships are mobile, their operating environments vary widely with location.

2-4.1 Derivation of the Projected EME

White [1] has proposed a systematic approach to development of the shipboard EME. Although somewhat abstract, it well illustrates the complexities involved. He points out that definitions of the environment must begin at design conception and be repeatedly updated as the design evolves. We start with the premise that the EME is generated primarily by both friendly and hostile forces which the ship expects to encounter in carrying out its mission. It follows that the EME is therefore a function of the tactical and physical interaction of these forces. Consequently, it is the interaction of these forces that we must first examine.

The examination begins with analysis of the ship's tactical and operational objectives, and, based on the nature of friendly assets, mission doctrine, and hostile threats, a representative EME range is developed. From this, for each type of engagement, a set of worst case and best case values are derived to bound the EME in the ship design. During the early stages, it is recognized that there will be substantial uncertainties. However, the EME definitions will be cumulatively improved as the ship design progresses. Therefore, in order to remain the best bound of values, it is necessary to iterate the definitions along the way.

The uncertainties at the beginning are due in large part to lack of information about ownship contributions. It should be remembered that at the time of initial EME definitions, the ship is from seven to ten years away from delivery, and, therefore, not by any means in final configuration. The definitions improve as the equipment acquisition and ship design proceed and as measured data replace predictions.

White acknowledges that this process of EME definition is rather difficult. Nevertheless, he notes that information on friendly and hostile forces is available in a number of publications (e.g., [3]), and, though incomplete, hostile threat analyses are routinely updated from intelligence work. From such data it is possible to calculate the EME across a set of engagements. The procedure entails a step-by-step look at each engagement to determine the nature of EM emissions anticipated, whether intentional or unintentional, friendly or hostile. Added to this must be the combined onboard transmitter and receiver operating bands, radiation power levels and component sensitivities, spurious output levels, effects of the passive shipboard EME, and susceptibility characteristics of ownship sensors. With so many variables, the result is, of course, only a best estimate, but still of great value in predicting an EME that is necessarily of the future.

2-4.2 EME Definition Guidance

There are certain positive steps that may be taken in developing a definition of the projected EME.[4] These include: (1) laboratory simulation of the anticipated environment through modeling and testing. A distinct merit of simulation techniques is that the models are easily updated and reused during progress of the ship design; (2) anechoic chamber measurements, where the models are subjected to testing in carefully controlled and shielded electromagnetic environments simulating the anticipated shipboard EME; and (3) full-scale measurements where the actual equipment and systems are tested in a full-scale representation (mockup) of the anticipated shipboard EME. This last method is quite costly but offers significant benefits in determining the performance of systems under "real-world" conditions. It has been used, in fact, by the Navy at so-called land based test facilities for recent-design warships, with excellent success.

Finally, in gathering the information useful to defining the shipboard EME, the following list is helpful:[5]

- What is the system intended to do?
- Is it tactical? Mobile? Transportable? Fixed plant? Strategic? Target-dependent?
- Does it stand alone, or is it part of a larger system?
- What are the signal inputs and outputs, and their range of frequency and power?
- What are the frequency management constraints and requirements?
- What are the basic power requirements?
- What are the range requirements?
- What is the sensitivity requirement for the receiving equipment?
- Where will the system be used?
- What will the platform structural environment be?
- Is the system required to operate continuously or intermittently?
- Are there any location, size, or weight restrictions?
- Is the system critical to a specific mission operation, and if so, what?

- Are there critical sequences of operations involving this system?
- To what extent will malfunction affect mission success or personnel safety?
- If antennas are involved, what special characteristics should be considered?
- Is the system active or passive (that is, does it transmit, receive, or both)?
- Is signal processing equipment required?
- With what equipment does the system interface?
- What modulation system will be used?
- What type of waveforms are involved?
- What sensitivity and resolution are required?
- What are the minimum threshold responses, both amplitude and duration?
- What are the accuracy requirements?
- Is this an analog or digital operation?
- Are there any special remote control requirements?
- In what type of facility is the equipment to be installed?
- What other equipment will be in the same installation?
- Are any inherent, definable problems expected?
- Are space-available problems anticipated?
- Are any special co-site problems anticipated?
- What are the inherent shielding characteristics of the installation?
- Will the system or equipment be exposed to enemy electronic countermeasures (ECM)?

The questions listed above are a good representation of the type of information needed. However, each ship is unique, and similar questions will have to be posed on a case-by-case basis before a clear definition of any one EME is derived.

REFERENCES

1. O. S. White, "EME Definition for Naval Vessels; Compliance with Requirements" (unpublished manuscript, circa 1983) p. 11.
2. *Electromagnetic Compatibility Management Guide*, MIL-HDBK-237A, Department of Defense, Washington, DC, 2 February 1981, p. 4.
3. *Electromagnetic (Radiated) Environment Considerations for Design and Procurement of Electrical and Electronic Equipment, Subsystems, and Systems*, MIL-HDBK-235-1A, Department of Defense, Washington, DC, 5 February 1979, pp. 7–9.
4. *Ibid.*, p. 10.
5. MIL-HDBK-237A, *op. cit.*, p. 25.

Chapter 3
Shipboard Electromagnetic Compatibility (EMC)

3-0 DEFINING EMC

One of the more formal definitions of EMC is: "the capability of electronic and electrical systems, subsystems, equipments, and devices to operate in their intended operational environment at design levels of performance and safety without suffering or causing unacceptable degradation because of unintentional electromagnetic radiation."[1]

Another of the official versions is: EMC is the ability of electronics equipment or systems to operate in a fixed environment within design levels of performance without degradation due to electromagnetic interference.[2]

Keiser perhaps states it best by stripping away all bureaucratic puffiness and offering simply: "EMC is that happy situation in which systems work as intended, both within themselves and in their environment."[3]

No matter which definition one might prefer, it should be well apparent from our examination of the subject in the previous two chapters that the "happy situation" of achieving total EMC in a surface ship is a fantasy. The record file of some 6,000 documented cases of shipboard EMI is evidence enough of reality. Yet, improving shipboard EMC is a necessary goal. In fact it is a requirement.

> Electromagnetic compatibility considerations are mandatory throughout the Department of the Navy and will be applied in the research, design, development, production, procurement, installation, and operational use of equipments and systems. Each command, activity, project, or program office, laboratory, and facility within the Department of the Navy is responsible for the application and enforcement of EMC requirements and for the achievement of EMC within its respective area of cognizance.[4]

> Management and engineering personnel must establish and implement a procedure for integrating EMC engineering into all phases of the life cycle for ships, systems, and equipments.[5]

Accordingly, ship system EMC specialists have accepted the challenge and continue to strive diligently toward that goal.

3-1 IMPLEMENTING EMC MEASURES

Whether a new piece of equipment requires shipboard integration or the ship itself is being newly designed, EMC management engineering must begin early, in the very concept phases of design. Moreover, management control must continue throughout the whole project, through installation or construction and through the active life of the ship. Consequently, solid EMC planning must be formulated from inception.

The first step to be taken in such a process is to identify the intended shipboard environment thoroughly. As discussed in Chapter 2, this requires a synthesis of all expected electromagnetic emitters and receptors, a prediction of potential EMI sources and victims, and an analysis of both friendly and hostile threats. After the operational environment is established, careful design procedures ensure that the equipment or system will be integrated compatibly into the environment to support the ship mission effectively. That is, EMC measures must provide a high probability of the system's being not only compatible within itself, but, just as importantly, within the overall ship environment. To realize that objective, the project development generally includes such engineering processes as modeling, simulating, testing, and analyzing to determine radiation and susceptibility characteristics and environmental constraints. To carry out this process properly, an effective EMC management plan is required.[6]

3-1.1 The EMC Program Plan

To achieve optimum electromagnetic compatibility in the development and shipboard integration of a piece of equipment or a system, it is imperative that the project manager prepare a detailed EMCPP. Such a plan is the principal management engineering document to be followed during each phase of the design. It outlines naval EMC policy, design philosophy, and organizational responsibility, and it provides clearly defined guidance, task assignments, and milestones needed during the process to ensure EMC.

At initiation of the plan, authorization must be requested to allocate a portion of the frequency spectrum to the system under development.

3-1.1.1 Frequency Spectrum Management

Intensive, worldwide competition for use of the crowded frequency spectrum is one of the chief causes of EMC's being of such interest to modern naval ship designers. Because of the competitive needs and the natural spectrum limits, proposed new frequency usages must undergo careful scrutiny in accordance with strict international regulations. Therefore, when commencing new equipment or system design, the project manager must submit a form DD 1494 requesting approval of a frequency allocation. The immediate purpose of this requirement is to enable the planned system to be designed without disrupting, or being disrupted by, other systems occupying that frequency or one nearby in the spectrum. Hence, the request is carefully reviewed for conformance to the spectrum utilization criteria of international and national regulatory bodies and of the Department of Defense.

Application for frequency allocation normally is requested at each of four stages in the system development:[7]

a. *Concept Development*—Allocation is required early in the concept phase, prior to the funding of studies or the fabricating of equipment test beds, even though little more than the system purpose, planned frequency band, and expected power output levels are known.
b. *Concept Validation*—A so-called experimental allocation is required before test model units are allowed to radiate electromagnetic energy, even when being tested in a controlled laboratory environment.
c. *Advanced Development*—Before a contract is settled for engineering development models, a third-stage allocation must be submitted showing measured test data (or calculated data when measured results are not available).
d. *Operational (Production) Development*—Prior to the signing of the contract for production units, an operational frequency allocation request containing technical characteristics and measured data is required.

For Navy purposes, the procedures for submitting a DD 1494 allocation request are given in OPNAVINST 2410.11. Spectrum management and related EMC policies within the Department of Defense are the responsibility of the Assistant Secretary of Defense for Command, Control, Communications, and Intelligence. Evaluation assistance in these matters is available from the Electromagnetic Compatibility Analysis Center in Annapolis, Maryland, where electromagnetic environmental data and equipment spectral characteristics are stored.

It should be emphasized here that a frequency allocation approval only allows the development to proceed in anticipation of future use of particular discrete frequencies or a frequency band. The allocation approval does not authorize *operation* of the equipment or system on the allocated frequencies. For this, a frequency assignment must be requested by the project manager and approval be granted before the system is put into operation.

Assuming the application for frequency allocation has been submitted properly, the EMCPP provides for adequate funding of the EMC effort and for the establishment of a panel of expert EMC advisors.

3-1.1.2 The EMCAB

Shipboard electromagnetic compatibility is so complex and multifaceted that it would be hopeless for the project manager, no matter how well-versed, to tackle the effort alone. Thus it is naval engineering policy to have an EMCAB serve the project in a technical advisory role. EMCABs are required in all new ship design programs, for major ship alteration projects, and for the development and purchasing of major equipment and systems. Members of the EMCAB are highly experienced specialists in shipboard EMC technology. They are appointed from within the naval system commands, from naval laboratories, and from private industry. Their purpose is to support the project manager in all aspects of EMC and systems performance by assisting with: (1) preparation of equipment and ship design specifications; (2) formal design reviews; (3) systems design analysis and predictions; (4) review of test plans and evaluation of test results; (5) development of systems installation criteria; and (6) ship construction and acceptance trials. In so doing, the EMCAB makes certain that EMC concerns are properly identified and that methods are employed throughout the program to achieve compatibility by adequately controlling EMI and precluding radiation hazards to personnel, fuel, and ordnance.

It is important that an EMC representative from the equipment manufacturer, or, in the case of new ship design, from the shipbuilding contractor, be assigned as a member of the EMCAB. In this manner the EMCAB is kept abreast of problems in the design and so can offer proposed tests and solutions in timely fashion. Conversely, the EMCAB is thus able to verify that the manufacturer or the shipbuilder is invoking naval specifications and criteria to ensure EMC during the design.

Another very important consultative function of the EMCAB is to appraise submitted ECPs with appropriate EMC advice and recommendations to the project Change Control Board (CCB). In its evaluation, the EMCAB differentiates the ECPs as follows:

a. *Category I—No Anticipated Problems*. ECPs in this category are checked against the Navy's EMI lessons-learned data base to determine whether similar programs under similar circumstances have experienced EMC problems. Contact is maintained with naval laboratories, engineering centers, and facilities to ensure concurrence in an evaluation of "no anticipated problems." For this category of ECP, the EMCAB prepares an EMC evaluation endorsement to be sent to the project leader stating, essentially, "A check with lessons-learned data base and cognizant engineering expertise indicates no anticipated EMC problems as of the date of this endorsement."

b. *Category II—Correctable Problems Anticipated.* ECPs in this category consist of those wherein specific and selective corrective action taken in the past has solved, or minimized, EMC problems which have arisen in conjunction with similar projects under similar circumstances. The EMCAB coordinates the inputs from naval laboratory and engineering activities pertaining to specific preventive or corrective measures taken. The EMC evaluation endorsement on these ECPs contains a short appraisal of the necessary specifics to be considered, together with documented data from lessons-learned or EM performance prediction analyses concerning the magnitude of problems that might be anticipated if specific corrective measures are not implemented.

c. *Category III—Severe Problems Anticipated.* ECPs in this category consist of those wherein substantial or severe EMC problems are anticipated if conditions are not corrected. In these instances, the anticipated problems are so complex, or the operational impacts so severe, that a formal EMC-impact engineering analysis of alternatives must be undertaken. The EMC evaluation endorsement regarding these ECPs must provide substantiated data pertaining to the anticipated risks of proceeding with the ECP as written, together with recommendations concerning the laboratories or engineering facilities most qualified to participate in the formal EMC engineering analysis. EMCAB responsibilities in this category include:

   1. Providing assistance in the preparation of necessary tasking documents.
   2. Giving advice on the estimated costs for the EMC impact engineering analysis.
   3. Monitoring the progress of the EMC impact analysis.
   4. Preparing recommendations from the engineering analysis report.

Additionally, the EMCAB must develop an EMI matrix as illustrated in Figure 3-1. The EMI matrix, showing the potential sources and victims of EMI, is the baseline for problem solving and analysis efforts by the EMCAB. As such the EMCAB develops plans of action to prevent or to correct EMI problems during the systems design and integration or ship alteration and systems installation processes. It is readily apparent, therefore, that the EMCAB is indispensable to the project manager in the continual striving for EMC.

SOURCE \ VICTIM	IFF	TACAN	Surface Search Radar	Air Search Radar	UHF LOS COMM	UHF SATCOM	BRDG-BRDG COMM	SHIP-AIR COMM	Fleet Broadcast	VHF Transceivers	SSES Receive	SATNAV	OMEGA NAV	HF Receiver	HF Transmitters
IFF	•	•				•									
TACAN	•														
Surface Search Radar															
Air Search Radar	•	•	•			•									
UHF LOS COMM					•	•			•			•			
UHF SATCOM					•	•									
BRDG-BRDG COMM								•							
SHIP-AIR COM							•								
VHF Transceivers										•					
HF Transmitters									•		•			•	•

Figure 3-1 EMI Matrix

3-1.1.3 The EMI Control Plan

A further requirement of benefit to the EMCPP, in which the assistance of the EMCAB is greatly valued, is the preparation of an EMI Control Plan. As part of the overall project development it is the responsibility of the equipment manufacturer, or, in the event of ship construction, the shipbuilding contractor, to submit a detailed methodology for meeting all contractual EMC requirements and environmental effects throughout the program. This is accomplished via the EMI Control Plan, which describes specific practices to be followed for separation and routing of cables and waveguides; application of grounding, bonding, and shielding techniques; prevention of electromagnetic safety hazards; ensuring EMC through quality control inspections; and conduct of test programs to demonstrate adequate suppression of EMI.

Technical requirements to be met in preparing the Control Plan are included in such documents as: MIL-STD-461B, *Electromagnetic Emission and Susceptibility Requirements for the Control of EMI*; MIL-STD-462, *Measurement of EMI Characteristics*; and MIL-STD-1310E, *Shipboard Bonding, Grounding, and Other Techniques for EMC and Safety*.

So from the start of the project to delivery and installation of the product, the EMCAB facilitates coordination between the manufacturer and the project leader by review and evaluation of EMI control measures needed to ensure EMC.

3-1.2 EMC Test and Evaluation

If a high degree of shipboard EMC integrity, once attained, is to be preserved, then equipment or systems which might dilute that integrity must be excluded from integration until made compatible. Likewise, so as not to degrade its own performance, any newly proposed system to be installed aboard ship has to be immune to the detrimental effects of the intended operational environment. Therefore, how can any assurance be established that: (1) a proposed new system will not dilute the EMC integrity of the host platform, and (2) the proposed new system will be so well hardened that it will perform as expected despite the harsh shipboard EME? The answer comes through the institution of a thorough test and evaluation program from inception to operational service. Such a program is essential to determine that the new system complies with the EMC provisions of the contract. Thus, the objective of the EMC test and evaluation program is to provide a high degree of confidence that the proposed system and its components will function according to specifications in the intended operational environment.

To carry out such a program, measurements are made in accordance with established standards and an approved TEMP. The resulting test data are analyzed and evaluated, and steps are taken to correct any deficiencies or failures. There are four major phases of EMC testing:

a. *During Concept Development*—To support early decisions on whether or not to proceed with the system design.
b. *During Design Validation*—To identify any design risks and to provide acceptable solutions.
c. *During Full-Scale Development*—To demonstrate that the design meets specified performance in the anticipated shipboard EME.
d. *After First-Article Production*—To correct deficiencies revealed during the operational evaluation.

Because of its importance in providing the basis for key design decisions, the TEMP is reviewed periodically for assurance that the test and evaluation is comprehensive and remains valid as the design develops and changes. The plan includes means of meeting the EMC requirements of appropriate standards, e.g., MIL-STD-449, *Measurement of Radio Frequency Spectrum Characteristics*; MIL-STD-461, *Electromagnetic Emission and Susceptibility Requirements for the Control of EMI*; MIL-STD-462, *Measurement of Electromagnetic Interference Characteristics*; and MIL-STD-469, *Radar Engineering EMC Design Require-*

ments. Also given in the plan are measurement objectives, test equipment configurations, test points, details of measurement procedures, and data recording format. It is required, too, that the test procedures be described in sufficient detail to enable the Navy project leader to have any of the testing duplicated for further analysis.

To be completely viable, operational testing of the newly developed system should be conducted in the most realistic shipboard EME possible. That is, if the equipment or system is to be placed where it will be subjected to high levels of electromagnetic energy, such as in a ship topside, then tests should be performed to verify satisfactory operation in the intended environment. This would include normal simultaneous operation of all shipboard emitters and sensors, and making use of data acquired from previous electromagnetic environmental predictions and operational experience. Finally, requests for approval of service use of the new equipment or system must include certification that the requisite EMC (self and platform) has been achieved.

3-1.3 EMC Configuration Management

It would be hopeless to expect that a satisfactory level of shipboard EMC could be preserved unless complete control of the shipboard configuration itself were strictly maintained. Hence, effectual configuration management is a requirement in all naval EMC programs, whether for new ship design or for major modifications and alterations. Even so, actual experience has shown that frequently there are wide variations in the same system installed in the same class or type of ship. Despite all the efforts expended on analytical and modeling techniques during the design phase to establish the appropriate baseline configuration, in practice changes still too often have been approved and incorporated without proper evaluation of the effect of EMC. Unfortunately, these variations frequently result in degraded performance of the installed system and, therefore, of the ship mission.[8]

A case in point is in the engineering design of shipboard topside arrangements, where the primary objective is to provide optimum coverage and performance of guns, missile launchers, weapon directors, radars, and communication systems to fulfill the ship's warfighting mission. This objective is very difficult to attain during new ship design, and it is even more difficult to preserve throughout the active lifetime of the ship because of the continual process of modifications and alterations. Consequently, to maintain good EMC, it is crucial that any proposed changes to the ship configuration be carefully evaluated to ascertain the extent of EMC effect. Such evaluations have to be completed in sufficient time to decide whether the change should in fact be allowed, or just what corrections are needed in order not to disturb the EMC.

One effective means of maintaining shipboard configuration management is to require that changes be made only through formal approval of an ECP or a SHIPALT. In this process it is incumbent upon the project manager to make sure that appropriate EMC analyses are conducted and that an EMC Impact Statement be included in the ECP or SHIPALT. Information in the Statement should include any proposed changes in the physical location of equipment; changes in the emission characteristics (e.g., frequency, modulation, power output, and antenna type); changes in sensor characteristics (e.g., bandwidth, sensitivity, selectivity, filtering, frequency, and antenna type); or changes to the ship hull structure which could affect shielding, bonding, and grounding integrity. Moreover, the statement must contain supporting rationale for the originator's proposed changes.

In many instances a simple review of file case studies will aid in predicting whether similar system changes in similar situations have caused any problems. In such events EMC troubles can be anticipated and prevented by applying known solutions. But in other cases the system integration problems resulting from alterations can be so complex, and the effects expected so detrimental, that an in-depth EMC evaluation is quite necessary. In such cases the project manager must assess the risks involved as well as the results of not adopting corrective measures. Here again, the EMCAB is heavily relied upon to provide advice of critical importance to preserving the well-being of the program.

3-1.4 EMC Training and Awareness

We turn our attention now to yet another facet of shipboard EMC by no means of least importance. The subject is training. For no matter how earnestly the engineering designer and program manager have worked together to produce a unit of equipment or a system or a new warship having an optimum initial level of EMC, the product *user* must be fully aware of the need for constant EMC upkeep. Left unattended, shipboard electromagnetic systems and compatibility will degenerate inexorably due to the natural consequences of time and change. Thus, an appreciation of shipboard EMC and the deleterious consequences of incompatibility must be made known to the operators and users. It behooves the project manager, therefore, to prepare an EMC training plan whenever newly developed equipment is to be introduced, or a system modified, or a new ship delivered. The plan should address in particular the procedures required for preserving the total platform EMC.

Part of the problem, even among ship personnel trained in the operation and maintenance of electronic equipment, is that there may likely be a general lack of knowledge of the many causes of EMI.[9] Ship operators quite often may be unaware of the electromagnetic subtleties which work insidiously to degrade system performance. Even routine maintenance procedures required for a continued high degree of individual system performance can be damaging to the total ship EMC. This is especially true in the case of ostensible improvements which involve modifying or altering a topside system. Such parochial changes can cause serious overall system degradation. Therefore, it is very important that each new installation or modification be thoroughly assessed and tested for full compatibility in the shipboard EME.

Ship personnel should be aware also that many EMI problems can be avoided by such everyday practical techniques as proper tuning and aligning of electronic equipment; careful bonding, grounding, and stowage of topside items; operating transmitters within prescribed power limits and with adequate frequency spacing; and selection of alternative antennas for communications circuits. Likewise, operators should be conscious of the many electromagnetic susceptibility mechanisms that contribute to upsetting the delicate balance of EMC. Furthermore, ship technicians and operators should be trained to identify sources of performance degradation and taught how to employ EMI reduction methods to restore good performance.

Recognizing this vital need for EMC awareness, the Navy now requires each ship to have an EMI control officer assigned the responsibility for maintaining the ship's EMC integrity. As such, in implementing an effective shipwide EMC awareness program, the EMI control officer must:[10]

a. Develop and implement a training program to ensure that all crew members are kept informed of the need for shipboard EMC, and of what each individual is expected to do toward maintaining EMC.
b. Develop management and inspection procedures to ensure that all ship's force efforts are coordinated and scheduled to achieve, restore, and maintain EMC.
c. Ensure that proper corrective maintenance is performed on equipment or systems causing EMI.
d. Procure and maintain test equipment needed for EMC testing, and ensure that all equipment is properly calibrated.
e. Ensure that thorough and comprehensive inspecting, testing, grounding, and bonding techniques are practiced to detect and suppress EMI.

In summary, the achievement of satisfactory shipboard EMC and its preservation throughout the lifetime of the ship is an arduous task. It requires a continual effort of awareness and training, as well as alert response on the part of the ship operations and maintenance personnel. Timely identification of EMC problems and aggressive corrective action are essential to the proper functioning of the total ship electromagnetic system in effective fulfillment of mission objectives.

REFERENCES

1. *Electromagnetic Compatibility (EMC) within the Naval Sea Systems Command*, NAVSEAINST 2410.2, Department of the Navy, 13 January 1977, Enclosure (1), Definitions.
2. *Procedures for Conducting a Shipboard Electromagnetic Interference Survey*, MIL-STD-1605(SHIPS), Department of Defense, 20 April 1973, p. 1.
3. B. E. Keiser, *Principles of Electromagnetic Compatibility*, Artech House, Dedham, MA, 1979, p. 1.
4. *Electromagnetic Compatibility Program within the Department of the Navy*, OPNAVINST 2410.31D, Office of the Chief of Naval Operations, 6 August 1984, p. 3.
5. "Engineering Requirements for Electromagnetic Compatibility," NAVSEANOTE 2410, Department of the Navy, 8 March 1983, Enclosure (1), p. 3.
6. *Electromagnetic Compatibility Management Guide for Platforms, Systems, and Equipment*, MIL-HDBK-237A, Department of Defense, 2 February 1981, pp. 4–12.
7. *Ibid.*, pp. 77–78.
8. *Ibid.*, pp. 39–40.
9. "Shipboard EMC Management," *The Electronics Material Officer's Guide to Shipboard Electromagnetic Interference Control*, STD-407-5287556, Department of the Navy, August 1980, p. 3-1.
10. *Ibid.*, p. 3-3.

Chapter 4

Shipboard Electromagnetic Interference (EMI)

4-0 THE SHIPBOARD EMI PROBLEM

Without question the crux of shipboard EMC engineering technology is the prevention and control of EMI. It has been so, as we discovered in Chapter 1, since the origin of shipboard electronics. Yet, despite the most diligent management techniques during systems design and production; despite expert knowledge, experience, and appreciation of the shipboard EME; and despite the best of efforts in training and awareness, EMI is ever present aboard naval ships. Its presence is due to the mere nature of the shipboard environment, the density of complex, highly sophisticated electronic systems, and the extraordinary requirements of critical ship missions. As a consequence, each ship must be tested, evaluated, and treated for EMI on a case-by-case basis. Therefore we will now examine in detail the engineering practices for effective control of onboard EMI.

4-1 SOURCES OF SHIPBOARD EMI

EMI is defined as any electromagnetic disturbance which interrupts, obstructs, or otherwise degrades or limits the effective performance of electronic and electrical equipment.[1] As confined to and contained within the boundaries of a ship, this definition encompasses an astonishing number of possible sources, by far the most of which are quite unintentional. Occurring through both conduction and radiation paths, shipboard EMI generally is categorized as being either natural or man-made. Figure 4-1 illustrates the relative amplitude and spectrum of these EMI sources.

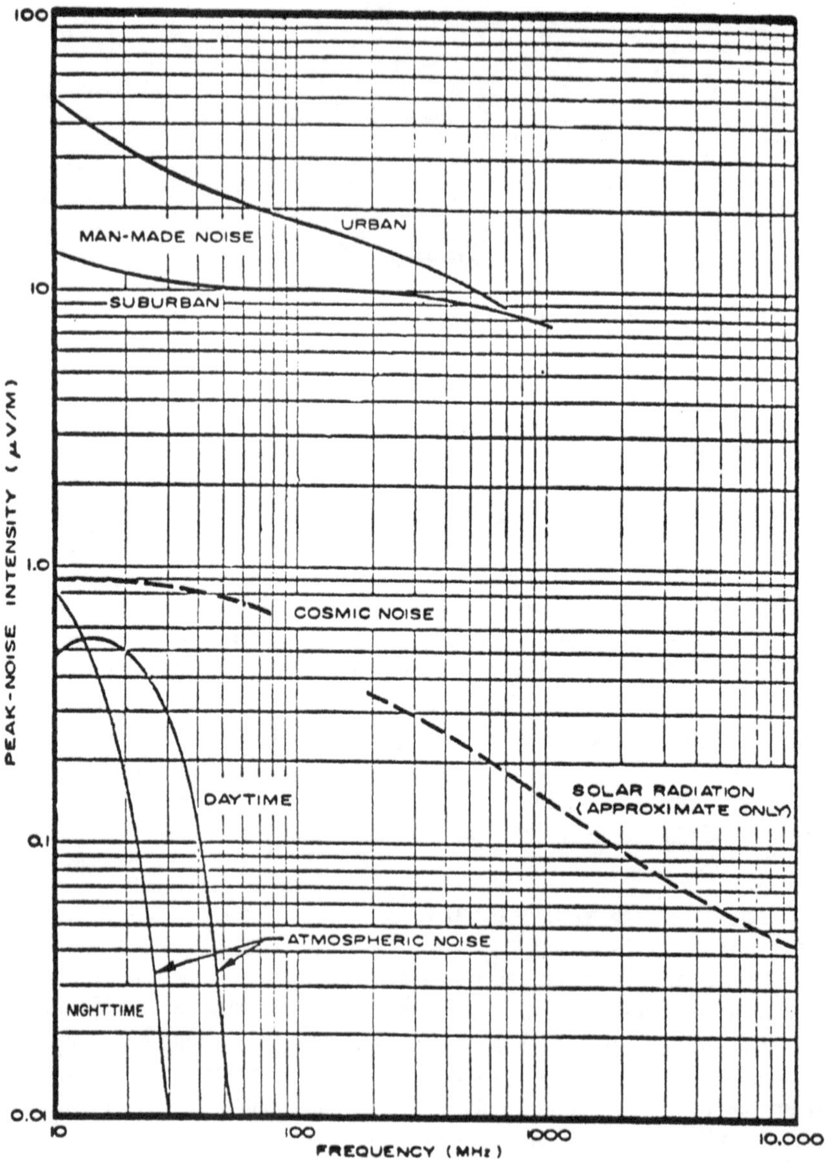

Figure 4-1 Sources of Electromagnetic Interference (from USAF EMC Handbook DH 1-4)

4-1.1 Natural Sources of Shipboard EMI

The world environment is replete with naturally occurring electromagnetic disturbances. These disturbances are created both within our earth's atmosphere and from a variety of points out in the universe. As a significant contributor to shipboard ambient noise levels, natural interference is a mixture of random discrete impulses and a steady broadband hiss. Fortunately, the characteristics of natural interference are well known and are reckoned with at the equipment design phases.

a. *Atmospheric Noice*—Generated primarily from lightning discharges, atmospheric noise produces intermittent high-intensity bursts of interference during local electrical storms and a continuous low-level rattling and crackling disturbance from numerous storms in the distance. This electrostatic inteference is strongest in tropical areas of high thunderstorm activity.

   As a natural phenomenon, atmospheric noise is present from very low frequencies to about 100 MHz. It is the predominant noise source, however, below 30 MHz, and, varying somewhat according to the season and whether it is night or day, it disturbs most strongly in the HF region, peaking at approximately 8 MHz.

b. *Cosmic Noise*—EMI originating in nature beyond the earth's atmosphere (i.e., in outer space) is classified as cosmic noise. This type of EMI is a combination of galactic, thermal, and interstellar noise emissions.[2] The amplitude of the composite cosmic noise is lower than that of atmospheric noise below 10 MHz. However, above 50 MHz, cosmic noise levels are notably higher than atmospheric noise. Moreover, cosmic noise is wideband, being bothersome in the VHF range, and annoying even out to EHF (well over 30 GHz). There are times, too, during cyclical sunspot activity, when cosmic noise bursts last several minutes and exceed atmospheric noise levels in the HF band.

   Two principal sources of noise interference within our solar system are: (1) nonthermal electron activity in Jupiter's magnetic field, and (2) thermal emissions from the moon caused by solar heating of the lunar surface. Outside the solar system the most intense noise source is the supernova star Cassiopeia A.[3]

4-1.2 Man-Made Sources of Shipboard EMI

Man, in contriving electrical apparatus to lighten his burdens and increase human comforts, has unwittingly allied with nature to produce even more sources of EMI. From rotating electrical machinery to electrical lighting to electromagnetic transmission of information, the byproducts of these benefits to mankind have been increased levels and varied types of electromagnetic noise. The more man-made noise is generated, the less reliable and less efficient become the electrical and electronic systems.[4] The problem is greatly exacerbated aboard ship, where very many electromagnetic devices and systems are required to operate simultaneously in a confined volume.

a. *Shipboard Transmitter System EMI*—In carrying out its routine operational functions, a naval ship has a number of RF emitters in concurrent service. These include several HF, VHF, UHF, SHF, and EHF communications transmitters; air search, surface search, and navigation radars; TACAN and IFF transponders; weapons detection, acquisition, and tracking directors; meteorological and telemetry data transmitters; and, at times, electronic warfare countermeasure emitters. Many of these radiate very high power, and some transmit omnidirectionally or rotate 360° continuously. As a consequence, the systems are capable of mutually interfering with each other. Furthermore, the onboard associated sensors are prone to intercept undesired emissions either through direct coupling or by multipath reflections. Even if the emitters are carefully designed so as to radiate intentionally only a specific frequency or band, in actual use unwanted RF energy escapes at a large number of spurious frequencies to cause potential EMI problems. Likewise, those emitters that employ highly directional antennas emanate undesired energy in the sidelobes and backlobe portions of the radiation pattern.

    Assuming that the transmitter systems designers have incorporated adequate interference suppression in the design and production of shipboard equipment, the following transmitter-related EMI spurious emission problems are frequently experienced aboard ship as a result of improper installation, operation, and maintenance:

1. *Harmonic Frequency Products*—Generated by nonlinearities in transmitter power output stages, harmonic products are integral multiples of the desired fundamental radiation frequency. Even though equipment manufacturers are required to design transmitters with second-order harmonics suppressed to 60 dB below the fundamental, and higher order harmonics 80 dB below, improper tuning and operation of a transmitter will likely result in production of harmonic interference by the forcing of nonlinear excitation.
2. *Cross-Modulation and Intermodulation*—EMI in these instances is caused by interaction of two or more signals present at the same time in a nonlinear circuit. Intermodulation results from the mixing (heterodyning) of signals to produce new frequency components, while cross-modulation occurs upon the transfer of modulation energy from one RF carrier to another. These problems often happen when antennas are

installed so closely to one another that energy is directly coupled between them and thereby fed across to the transmitter output stages to mix with the desired radiated signals.

3. *Parasitic Oscillations*—Parasitic EMI results from self-excitation of transmitter circuitry, causing oscillatory radiation of undesired energy. This problem usually occurs when incorrect alignment procedures are used, or upon physically disturbing the original circuitry (moving of wires and components) during troubleshooting and repair. It is extremely important to exercise care in replacing electronic parts with exact types to restore the precise configuration as originally installed.
4. *Sideband Splatter*—Spurious sideband components produced outside the intended modulated RF bandwidth result in an EMI known as splatter. Again, it usually results from faulty transmitter operation, either through overmodulation or through poor tuning practices such as overdriving the intermediate and final output stages by overzealous attempts to eke out the peak radiated power.
5. *Broadband Arcing Noise*—High power transmitters produce very high RF currents and voltages along the transmission system. If the transmitters are not properly matched and loaded into the antenna (maximum power transfer), standing waves along the transmission line can cause arcing and corona discharge. Similarly, RF energy induced in nearby rigging and structural appendages may exhibit arcing and sparking. The result is broadband noise.
6. *Waveguide and Coaxial Cable Leakage*—When RF energy escapes from poorly designed, installed, or maintained transmission lines such as waveguide and coaxial cable connectors and joints, undesired EMI is evidenced. This problem is particularly apparent in shipboard radar and microwave systems.

b. *Shipboard Receiver System EMI*—Although certainly not contributing such high levels of EMI as transmitters do, interference generated *within* a shipboard receiver may still have as pronounced an effect and result in serious performance degradation. Sources of internal receiver EMI (i.e., EMI originating within the receiver) include: (1) image frequency interference created by ordinary local oscillator heterodyne mixing but escaping unattenuated in a well-designed and filtered receiver because of faulty alignment and tuning; (2) extraneous interference signals produced in the receiver by intrusion of strong external signals coupled from nearby high power transmitters; (3) intermodulation and spurious interference products resulting from unintentional signal mixing in nonlinear receiver circuitry; and (4) cross-modulation when signals are unintentionally transferred from an undesired RF carrier to the intended receiver carrier.

c. *Shipboard Electrical Apparatus EMI*—A ship is in a sense a small, self-contained community. That is, in addition to providing a workplace and job for each onboard resident, it also supplies many amenities for comfort and entertainment: berthing, food, medical attention, sanitary facilities, and choices of leisure activities. A visitor on a guided tour aboard a modern naval warship might be surprised to see a barber shop, laundry, post office, variety store, library, pharmacy, clinic, carpenter shop, machine shop, gym, radio, TV, nightly movies, and even a brig.

What is not apparent to the casual observer is that the smooth operation of each of these facilities, in addition to the primary mission functions of the ship, is dependent upon a great number and variety of electrical apparatuses. These devices range from the smallest hand-held hair dryers to circuit breakers, switches, relays, massive propulsion system generators, large welding machines, and assorted lighting requirements throughout the ship. Each electrical apparatus is a potential source of undesired noise emissions adding to the ambient EMI level.

1. *Motors and Generators*—Broadband noise produced by shipboard motors and generators is a common but serious source of EMI. It is especially associated with arcing at the brush contacts of commutators and slip-rings. It also results from the instantaneous buildup and collapsing (current reversals) of electric fields and from frictional static discharges in belts, gears, and bearings. Additionally, harmonic components are generated in armature magnetic field nonlinearities.
2. *Circuit Breakers, Switches, and Relays*—The sudden opening and closing (so-called making and breaking) of electrical contacts results in both radiated and conducted wideband EMI. The usual occurrence is a voltage impulse transient as the circuit current is abruptly changed, causing an arc as the dielectric breakdown strength is exceeded between the metallic contacts. The noise spectrum for contact EMI ranges from VLF through UHF (about 10 kHz to 400 MHz).
3. *Engine Ignition Noise*—Ignition systems are commonplace aboard ship for use in such items as portable firefighting pumps and for starting the engines of helos and aircraft. These devices are perhaps the strongest source of man-made noise interference in the HF to VHF range (10 MHz to 100 MHz).
4. *Lighting*—Fluorescent lighting is employed throughout the internal spaces of a ship and is a notorious source of noise. EMI is created within lamps upon electrical breakdown. It is also conducted through the power circuitry and, most significantly, radiated from the power source connection lines.[5] This type of interference is troublesome from approximately 100 kHz to 3 MHz.

In addition to fluorescent lights, many ships use sodium vapor and mercury arc lamps for lighting the topside areas. Similarly, these lights generate electrical noise from 100 kHz to about 1MHz.

5. *Miscellaneous Electrical Items*—There are a great many sources of shipboard man-made EMI other than the major contributors listed above. These include such seemingly innocuous electrical apparatuses as heaters, power supplies, dc rectifiers, solenoids, rheostats, transformers, buzzers, PA systems, walkie-talkie radios, tape recorders, computers, data processing equipment, and microwave ovens. Each is a potential contributor to the overall shipboard noise.

d. *Hull-Generated Intermodulation*—Noise interference resulting from hull-related intermodulation is one of the more pronounced and widespread of shipboard EMI problems. It is man-made insofar as man provides the mechanism for its genesis. Yet the effect is natural; that is, it is evidence of nature taking its course. Hull-generated EMI, therefore, may be thought of as a hybrid interference. That it is promoted by the complex metallic structure of the ship and the harsh maritime operating environment cannot be denied.

As a shipboard electromagnetic phenomenon, hull-generated intermodulation is a direct consequence of: (1) the quantity of onboard transmitters and their radiated power levels; (2) the quantity of onboard receivers and their sensitivity levels; (3) the quantity of onboard antennas and their constricted placement; (4) the quantity of onboard operational frequencies in a congested spectrum; and (5) the quantity of possible nonlinear elements and junctions in the structural makeup of the ship.

Hull-generated intermodulation is oftentimes referred to as the ''rusty-bolt effect.'' It originates at many of the nonlinear components or junctions that abound in naval surface ships. Indeed, there have been estimates that perhaps thousands of often obscure nonlinear elements exist in the topside of any given ship. Moreover, it should be pointed out that steel itself is intrinsically nonlinear. Nevertheless, the majority of nonlinearities acting to create shipboard noise intermodulation is due simply to metallic junctions exposed to the sea environment.[6]

Table 4-1 lists a variety of metals by their standing in what is known as the galvanic series.
Note that materials commonly used in ship construction, such as aluminum and steel, are near the top of the list. As a consequence, they are metals that are more easily corroded and are classed as being least noble.

If two metals are joined together, the farther they are apart in the galvanic series, the greater the likelihood of chemical reaction producing corrosion. So long as the two metals are clean, dry, and held tightly in contact, the impedance between them is virtually zero.

Table 4-1. Galvanic Series of Metals

Corroded End (anodic or less noble)
Magnesium
Magnesium Alloys
Zinc
Aluminum 1100
Cadmium
Aluminum 2017
Steel or Iron
Cast Iron
Chromium Iron (active)
Ni-Resist. Irons
18-8 Chromium-nickel-iron (active)
18-8-3 Cr-Ni-Mo-Fe (active)
Lead-Tin Solders
Lead
Tin
Nickel (active)
Inconel (active)
Hastelloy C (active)
Brasses
Copper
Bronzes
Copper Nickel Alloys
Monel
Silver Solder
Nickel (passive)
Inconel (passive)
Chromium Iron (passive)
Titanium
18-8 Chromium-nickel-iron (passive)
18-8-3 Cr-Ni-Mo-Fe (passive)
Hastelloy C (passive)
Silver
Graphite
Gold
Platinum
Protected End (cathodic, or more noble)

Upon exposure to moisture, however, unprotected joints begin immediate deterioration due to oxidation and corrosion. The junction impedance then increases, and a semiconductor device is formed. When an electrolyte is present, such as sea water, a simple battery cell evolves, electrolytic action develops, and an accelerated rate of corrosion occurs at the less noble (anode) metal. The result is termed a nonlinear junction. If RF energy from onboard transmitters impinges upon, or is induced across, the junction (which acts as a dc rectifier), intermodulation signals are produced to emanate as EMI. Ideally, of course, only the same or adjacent metals in the galvanic series should be mated together. Use of such dissimilar metals as steel bolts through brass flanges or aluminum clamps across copper piping should be strictly avoided. The problem aboard naval ships is that there are numerous instances of aluminum in contact with steel. Therefore, even though these two metals are close in the galvanic series, corrosion will develop rapidly.

There are types of corrosion other than galvanic that create nonlinear junctions:

1. *Fatigue Corrosion*—Results from repeated vibrations and bending, whereupon the outer protective film of a metallic surface is broken and the corrosion process begins.
2. *Crevice Corrosion*—Occurs when shipboard contaminants and moisture combine to penetrate and collect in seams and crevices for a sufficient period of time to start corroding.
3. *Stress Corrosion*—Occurs when a metal is stressed to the point that miniscule cracking allows moisture to enter and initiate corrosion.
4. *Welding Corrosion*—Results when the intense heat of welding causes changes in the molecular structure of one of the similar metals being joined so that it becomes, in effect, a dissimilar metal, and, in the presence of moisture, begins to corrode.

e. *Intermodulation Theory and the Ship Hull Environment*—Just as when the local oscillator output of a receiver is heterodyned with a selected incoming signal at the nonlinear mixing stage to produce, by intermodulation, the new intermediate frequency (plus several discarded sum and difference frequencies), so are intermodulation products created by the mixing of extraneous electromagnetic energy in certain nonlinear elements of the ship hull. The problem with the hull-generated intermodulation, however, is that these frequency products are always unintentional and very surely unwanted shipboard EMI.

Assume that an RF signal from an onboard transmitter radiating frequency F_1 is by chance applied across an electrically nonlinear element in the ship hull structure. Intermodulation action results, and the frequency spectrum generated by this nonlinear mixing will contain the original fundamental F, plus several other frequencies harmonically related to F_1. That is, there will be a second harmonic $2F_1$, third harmonic $3F_1$, fourth harmonic $4F_1$, and so on. If two such RF signals of nonharmonically related frequencies F_1 and F_2 from separate transmitters simultaneously excite a nonlinear element, the output spectrum will contain not only the direct harmonic frequencies $2F_1, 3F_1, 4F_1, \ldots$, and $2F_2, 3F_2, 4F_2, \ldots$, but also many new frequencies related to the two fundamentals; *viz*:

$F_1 \pm F_2$, known as second-order intermodulation products
$2F_1 \pm F_2$, known as third-order intermodulation products
$2F_1 \pm 2F_2$, fourth-order products
$3F_1 \pm 2F_2$, fifth-order, and so on

In such a manner an enormous number of intermodulation products are unwittingly generated in the ship environment.

The basic equations in intermodulation theory for this event are given by:[7]

$$R = MT_1 + NT_2$$

and

$$Q = |M| + |N|$$

where T_1, T_2 = transmitter RF carrier frequencies expressed in like units, and $T_1 < T_2$.

M, N = integers; i.e., zero, positive, or negative
R = the resultant intermodulation product interference frequency (in the same units as T_1 and T_2)
Q = the order of intermodulation product

Therefore, in the case of $3F_1 \pm 2F_2$ above, $|M| = 3$, $|N| = 2$, and the intermodulation product is fifth-order. For both $2F_1 + F_2$ and $F_1 \pm 2F_2$, the yield is a third-order product. Likewise, if a third shipboard transmitter participates in excitation of the same nonlinear element, then third-order products could result from $F_1 \pm F_2 \pm F_3$, and so on, summarized as follows:

FREQUENCY	PRODUCT ORDER	FREQUENCY	PRODUCT ORDER
$F_1 \pm F_2$	2	$3F_1 \pm F_2$	4
$2F_1 \pm F_2$	3	$F_1 \pm 3F_2$	4
$F_1 \pm 2F_2$	3	$3F_1 \pm 2F_2$	5
$2F_1 \pm 2F_2$	4	$2F_1 \pm 3F_2$	5

Figure 4-2 Intermodulation Product Orders

Figure 4-3 illustrates how rapidly the number of intermodulation products increase with an increasing number of RF exciters.

It can be seen that 10 transmitters simultaneously radiating discrete frequencies theoretically could produce 670 third-order intermodulation products and over 20,000,000 13th-order products! It should be pointed out that intermodulation products as high as the 60th order have been actually recorded during shipboard EMI tests.[8] Figure 4-4 depicts the dramatic effect of adding just one more transmitter. Here T is the number of transmitters in service, Q is the intermodulation product order, and Pa is the number of products generated as a function of increasing the number of radiated exciters. For example, if 12 transmitters are radiating energy that excites a nonlinear element, and a 13th transmitter is added, the result would be approximately 25 new second-order products, 300 third-order products, 2500 fourth-order, and about 12,000 fifth-order.

NO. OF TRANS-MITTERS	NUMBER OF ODD-ORDER PRODUCTS					
	3	5	7	9	11	13
1	1	1	1	1	1	1
2	6	10	14	18	22	26
3	19	51	99	163	243	339
4	44	180	476	996	1,804	2,964
5	85	501	1,765	4,645	10,165	19,605
6	146	1,182	5,418	17,718	46,530	104,910
7	231	2,471	14,407	57,799	180,775	474,215
8	344	4,712	34,232	166,344	614,680	1,866,280
9	489	8,361	74,313	432,073	1,871,845	6,539,625
10	670	14,002	149,830	1,030,490	5,188,590	20,758,530

Figure 4-3 Number of Possible Odd-Order Products Versus Transmitters Operating Simultaneously

The potential for hull-generated intermodulation interference aboard naval ships cannot be taken lightly. Degradation from such EMI, especially to shipboard communications, can be severe. Table 4-2 is a list of some of the representative items that act as nonlinear devices in the shipboard environment. Methods of dealing with the problem of hull-generated rusty-bolt-type intermodulation will be taken up in later sections of this chapter.

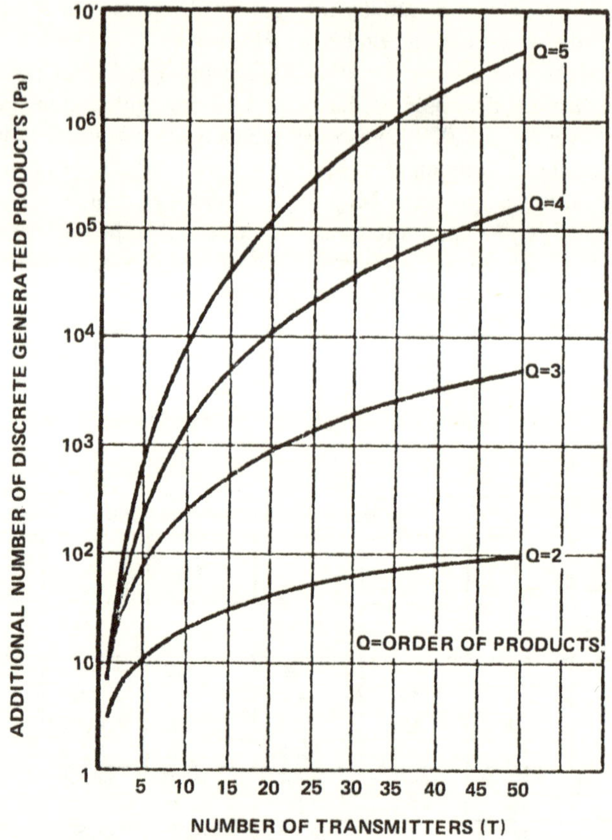

Figure 4-4 Intermodulation Product Increase

f. *Reflected Energy Multipath EMI*—Another source of shipboard EMI of major import is reflected electromagnetic energy. This type of interference is a consequence of the congested nature and peculiar structure of a ship. As we noted in the beginning of Chapter 2, the ship contains multiple scattering obstacles, including such items as superstructure geometry, deck houses, masts, yardarms, stanchions, booms, davits, weapon systems, lifelines, and a variety of large and small antennas. When RF energy radiated outward in a desired direction is inadvertently reflected from one

Table 4-2. Typical Ship Hull Nonlinear Junctions

Antenna pedestals	Hoist cables
Armored cable	Jackstaffs
Atomic fallout washdown systems	Ladders
Awning supports	Lifeboat holders
Belaying pins (signal flag ropes)	Life jacket holders
Boat cradles	Life raft holders and racks
Bolted flanges or panels	Masts
Bonding and grounding straps (deteriorated)	Radar waveguide flanges
Booms (refueling and loading)	Rigging
Cable clamps	Rusty or corroded bolts and screws
Cabinets	Safety nets
Canopy supports	Scuttles
Conduit	Shackles
Cranes	Stanchions
Davits	Storage racks and bins
Dissimilar metals	Swivels
Doors	Tackle
Drainpipes	Transmission line, circular
Expansion joints	Transmission line, rectangular
Fog nozzles	Turnbuckles
Gratings	Waveguide
Handrails	Wire mesh covers
Hatches	Yardarm rails
Hinges	

or more such obstacles in its path, it is likely to be coupled into highly sensitive receivers nearby. Serious degradation in sensor performance results. Therefore, careful topside design and configuration control must be exercised to preclude reflected energy EMI.

4-2 SHIPBOARD EMI CONTROL

The problem of shipboard EMI may be considered satisfactorily under control only when each of the ship electronics systems operates properly both independently and in concert with all other ship systems. That is, no individual system will act as a source of interference to affect the operation of any other system adversely, nor will any equipment or system be affected adversely by external sources of interfering electromagnetic energy. Therein would be the happy situation of total EMC referred to early in Chapter 3.

Such a utopian goal would be realized only if the most skillful ship and equipment technologists were fully able to design, develop, produce, install, and maintain their systems in such a way that all possible EMI sources were anticipated and eliminated. Because so ideal a goal is not realistically achievable, artful techniques must be practiced to restore and maintain EMC integrity. Those techniques include shielding, bonding, grounding, filtering, electronic blanking, and ship topside design.

4-2.1 Shielding Techniques

Stop it at the source. Contain it at the root. That should be the first maxim of good EMI control. If an electrical device or electronic system could be prevented absolutely from emitting RF disturbances, there would be little problem left to deal with. However, not all forms of undesirable emissions can be contained. For example, some energy will always radiate unintentionally from antenna sidelobes; some portion of desired radiation patterns will encounter topside obstacles and be scattered and reflected; and some nonlinear elements of the ship hull will be excited by induced RF currents to produce intermodulation interference. Therefore, the second EMI control maxim should be: Prevent it from entering. Keep it out.

In either event, whether to contain internally-generated EMI at its root, or to prevent external EMI from penetrating a potential victim, the technique most commonly applied is some form of adequate shielding.

4-2.1.1 Shielding Theory

In effect, as an electromagnetic barrier or protective shroud, shielding is a means of providing sufficient isolation between source and victim. Therefore, shielding may be thought of as a decoupling mechanism inserted to minimize mutual interaction. It is applied primarily to equipment and cables. Our interest here is in the use of EMI shielding techniques after systems integration into the ship; i.e., despite the best of EMI control practices (e.g., MIL-STD-461) during equipment design and manufacture to meet military requirements, and despite attempts to eliminate the interference by relocation or reorientation of the offending source or the victim.

When electromagnetic energy meets a metal shield it experiences an abrupt transmission mismatch due to a change in impedance. Depending upon the frequency (wavelength) of the energy relative to the type of metal and material thickness, the desired effect is to have some energy reflected at the point of entry, some attenuated and dissipated as heat by absorption through the material, and some reflected at the exit point. The quality or merit of the shield in satisfactorily reducing the energy is known as the shielding effectiveness, *SE*, expressed in dB:

$$SE = 20 \log_{10} \frac{E_1}{E_2} \text{ dB}$$

where

E_1 = field strength measured without shield
E_2 = field strength measured after shielding

In determining the actual material to be used and its wall thickness, the effectiveness equation used in design practice is:

$$SE = A + R + K$$

where (from [2]) A = the absorption, or penetration, loss calculated by

$$A = 1.314\, T\sqrt{F\sigma_r\mu_r} \text{ dB}$$

where

T = wall thickness in cm
F = frequency in hertz
σ_r = conductivity relative to copper
μ_r = permeability relative to copper

R = the reflection loss determined for far-field (free space 377 ohm impedance) plane wave radiated energy by

$$R_p = 168\text{–}10 \log_{10} \frac{F\mu_r}{\sigma_r} \text{ dB}$$

or near-field high impedance (electric component) energy by

$$R_E = 362\text{–}20 \log_{10} D \sqrt{\frac{\mu_r F^3}{\sigma_r}} \text{ dB}$$

where D = distance from radiating source to shield in cm, or near-field low impedance (magnetic component) energy by

$$R_H = 20 \log \left[\frac{1.173}{D} \sqrt{\frac{\mu_r}{F\sigma_r}} + 0.053 \sqrt{\frac{F\sigma_r}{\mu_r}} + 0.354 \right] \text{dB}$$

and K = a correction factor for multiple reflections occurring inside very thin-walled shields.

Table 4-3 (from [2]) provides the absorption loss, A, for various metals per millimeter thickness at 150 kHz.[9]

Table 4-3. Shielding Absorption Loss

Metal	*Relative Conductivity* σ_r	*Relative Permeability at 150 kHz* μ_r	*Absorption Loss A, dB/mm at 150 kHz*
Silver	1.05	1	52
Copper-Annealed	1.00	1	51
Copper-Hard Drawn	.97	1	50
Gold	.70	1	42
Aluminum	.61	1	40
Magnesium	.38	1	31
Zinc	.29	1	28
Brass	.26	1	26
Cadmium	.23	1	24
Nickel	.20	1	23
Phosphor-Bronze	.18	1	22
Iron	.17	1,000	650
Tin	.15	1	20
Steel, SAE 1045	.10	1,000	500
Beryllium	.10	1	16
Lead	.08	1	14
Hypernick	.06	80,000	3500*
Monel	.04	1	10
Mu-Metal	.03	80,000	2500*
Permalloy	.03	80,000	2500*
Steel, Stainless	.02	1,000	220*

*Assuming material not saturated

For those who prefer the use of nomographs Figures 4-5 through 4-7 (from [3]) allow easy determination of the reflection loss in the English system of units (separation distance, D, in inches, and shield thickness, T, in mils).

To find R_p (plane wave energy) from Figure 4-5:

a. Locate metal to be used on σ/μ scale
b. Place straightedge between metal and frequency of use on F scale
c. Read reflection loss in dB on R_p scale.

Example: For copper at 1 MHz, R_p = 108 dB.

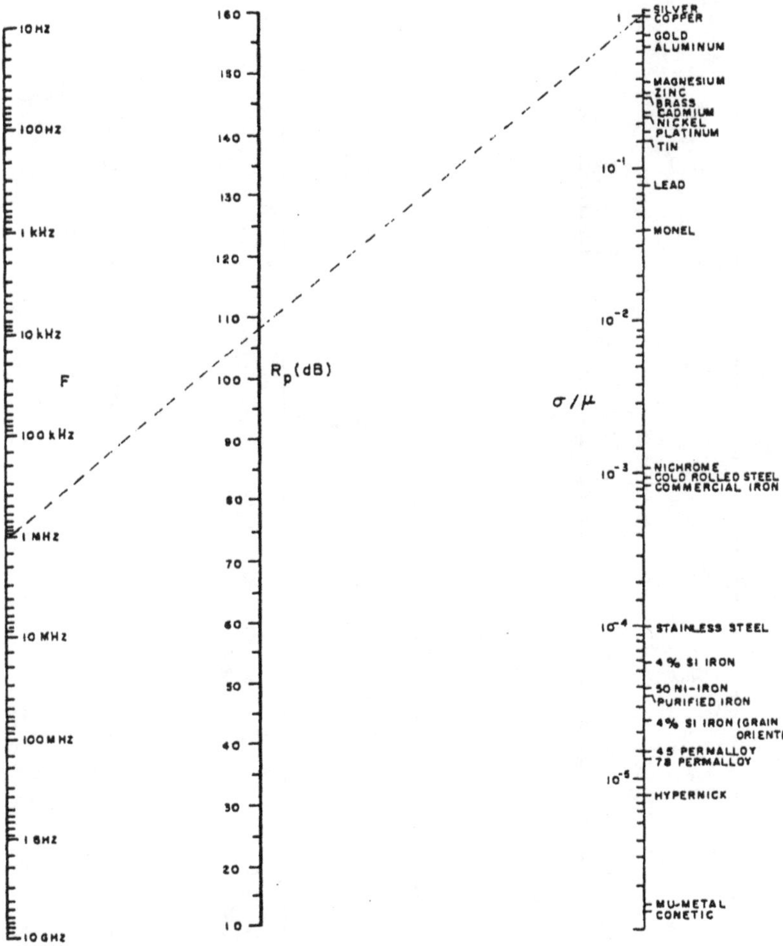

Figure 4-5 Plane Wave Reflection Losses R_p

To find R_E (high impedance E-field) from Figure 4-6:

a. Locate metal to be used on σ/μ scale
b. Place straightedge between metal used and distance from source to shield on D scale (in inches)
c. Mark point where line crosses blank scale
d. Place straightedge between point on blank scale and frequency of use on F scale
e. Read reflection loss in dB on R_E scale.

Example: For aluminum at 1 MHz with 12 inches separation between source and shield, R_E = 150 dB.

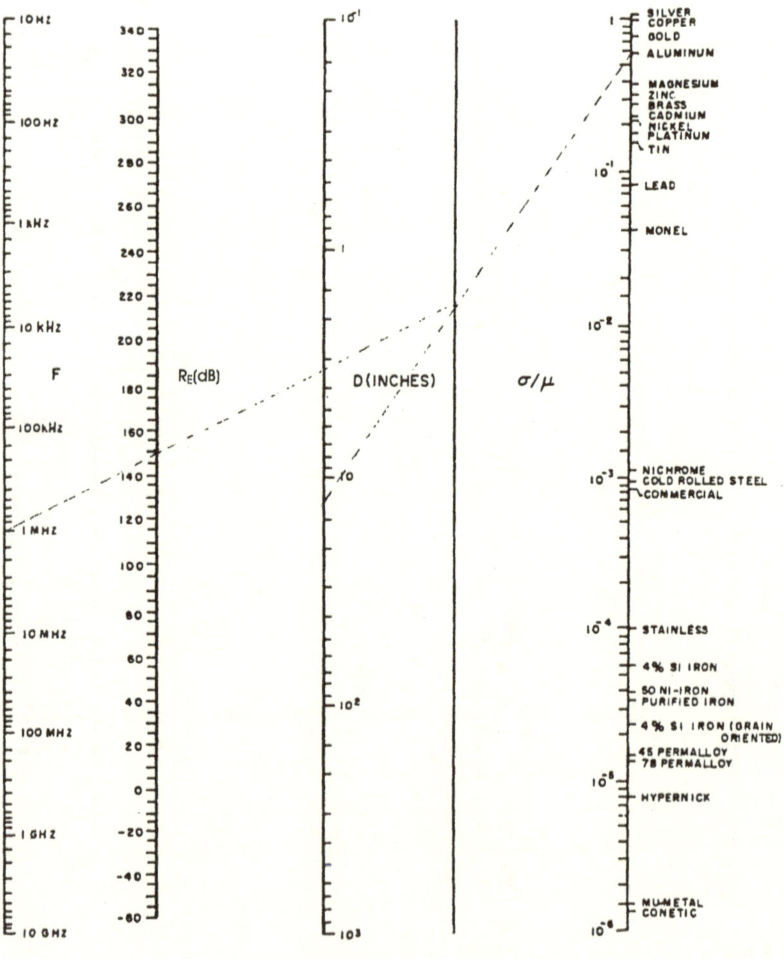

Figure 4-6 Electric Field Reflection Losses R_E

Similarly R_H (low impedance H-field) is obtained from Figure 4-7.

Example: For aluminum at 1 MHz with 12 inches separation from source to shield, $R_H = 62$ dB.

A comparison of reflection losses for copper, aluminum, and iron at various frequencies is illustrated in Tables 4-4 through 4-6.

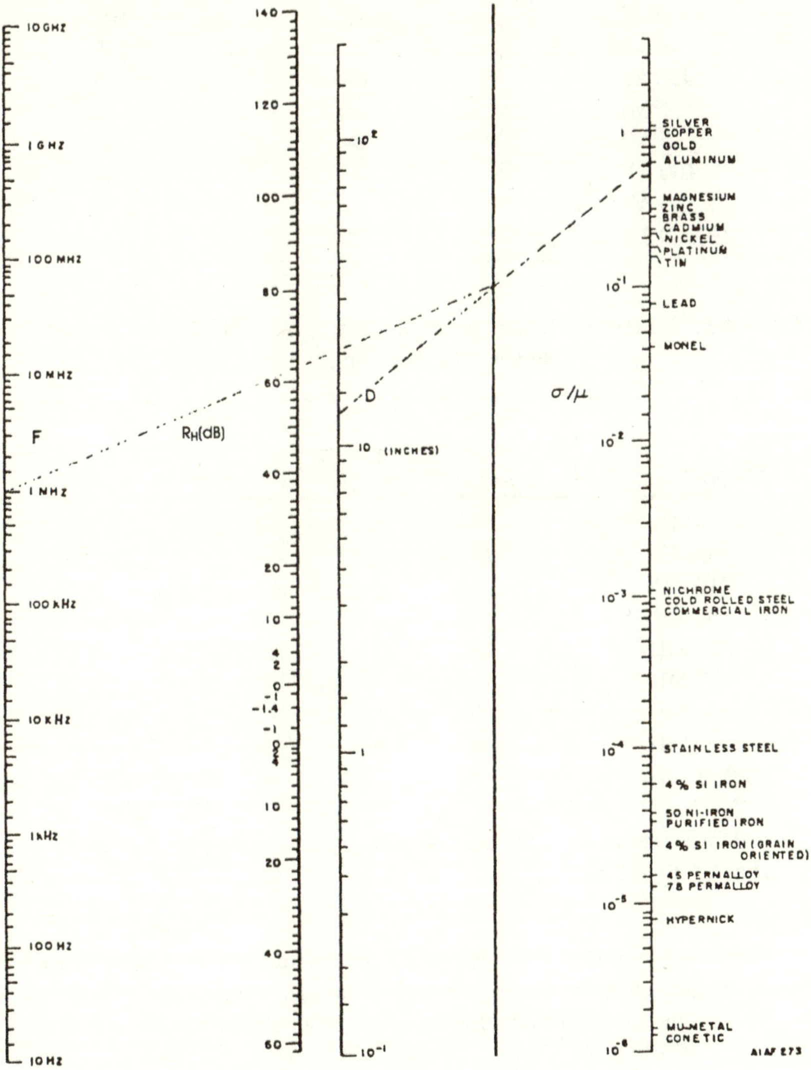

Figure 4-7 Magnetic Field Reflection Losses R_H

Table 4-4. Plane Wave Reflection Loss R_p (Far-Field Impedance is 377 Ohms)

Frequency	*(Loss in dB)*	
	Copper	*Iron*
60 Hz	150	113
1,000 Hz	138	100
10 kHz	128	90
150 kHz	117	79
1 MHz	108	72
15 MHz	96	63
100 MHz	88	60
1,500 MHz	76	57
10,000 MHz	68	60

Table 4-5. E-Field High Impedance Reflection Loss (12 Inches Separation Between Source and Shield)

Frequency	*dB Loss*		
	Copper	*Aluminum*	*Iron*
60 Hz	279	—	241
1000 Hz	242	—	204
10 kHz	212	—	174
150 kHz	177	175	—
1 MHz	152	150	116
15 MHz	117	115	83
100 MHz	92	90	64
1500 MHz	*	—	*
10,000 MHz	*	—	*

*At these frequencies, the fields approach plane waves with an impedance of 377 ohms

Figure 4-8 is useful for determining the shield thickness required for alternative metals when the desired absorption loss and frequency are known.

To find the material thickness required for an absorption loss of 10 dB at a frequency of 100 kHz:

a. Place a straightedge between 100 kHz on the frequency scale and 10 dB on the A scale

Table 4-6. H-Field Low Impedance Reflection Loss (12 Inches Separation Between Source and Shield)

Frequency	*dB Loss*		
	Copper	*Aluminum*	*Iron*
60 Hz	22	—	-1
1,000 Hz	34	—	10
10 kHz	44	—	8
150 kHz	56	54	19
1 MHz	64	62	28
15 MHz	76	74	42
100 MHz	84	82	56
1,500 MHz	*	—	*
10,000 MHz	*	—	*

*At these frequencies, the fields approach 377 ohms in impedance and become plane waves

b. Draw a line between F and A extending to intersect the blank scale
c. Pivot straightedge around point on blank scale to type metal desired
d. Draw line from metal through pivot point to point on thickness scale.

Example: Copper requires a shield thickness of 9.2 mils, and stainless steel requires a thickness of 5 mils.

For electrically thin shields having an absorption loss less than 10 dB, a correction factor, *K*, must be added to the shielding effectiveness equation. Figure 4-9 is a graph from which *K* may be found for varying thicknesses of copper from 10 Hz to 1 MHz. Table 4-7 gives values of *K* for copper and iron shields from 60 Hz to 1MHz for both the E-field and H-field interference sources.

In summary, shielding effectiveness is a combination of reflective and absorptive losses which result when RF energy encounters a metallic barrier. The characteristic difference between magnetic and electric fields should be borne in mind when selecting the type of metal and wall thickness to be used. For example, because reflection losses are small for magnetic vectors, good shielding effectiveness for H-fields requires high absorption-type losses. Also, at the lower frequencies, both the reflection and absorption losses are small for common metals such as copper and iron; therefore, special high-permeability magnetic alloys like nickel-iron Mumetal must be used for effective H-field shielding.

For the case of electric field vectors, reflection losses are higher than absorption losses, so E-field shielding is more easily achieved. Consequently, metals having high conductivity such as copper or aluminum are most often employed.

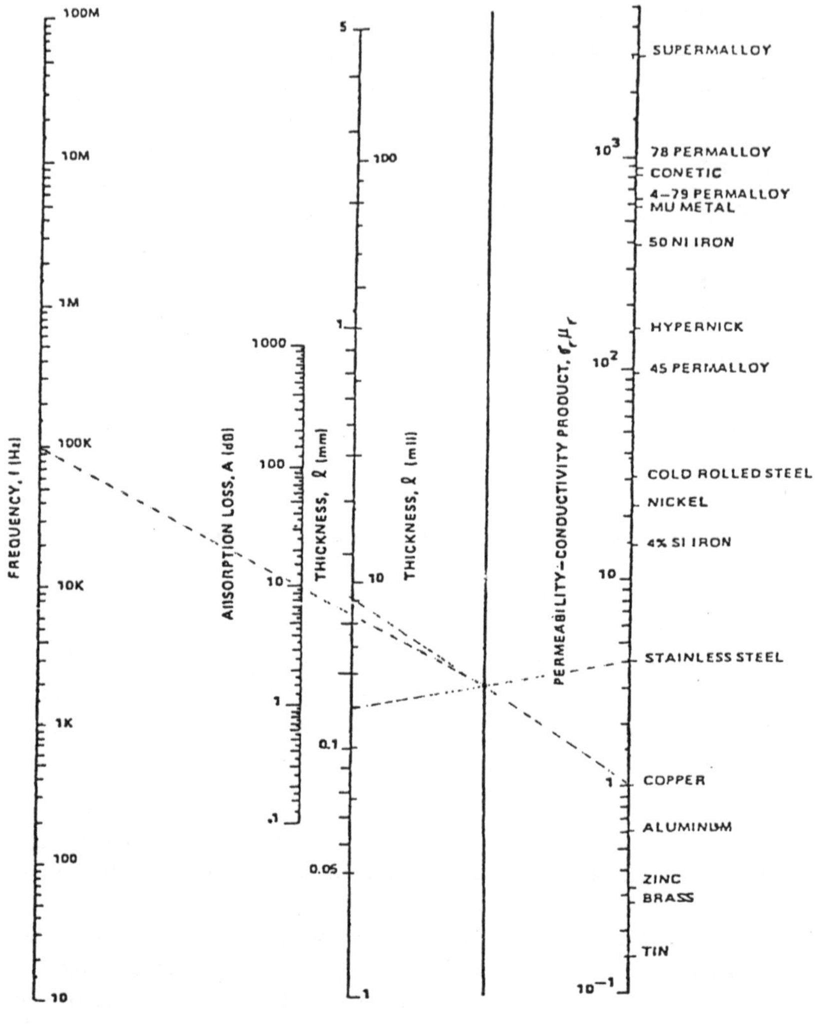

Figure 4-8 Shield Thickness

Finally, it should be noted that shielding effectiveness is a function of physical parameters, too; that is, the manner in which the shield is shaped and fastened in place. No shielding is absolutely perfect. Some energy will inevitably penetrate the barrier through seams, edges, cable entrances and access openings, and fastener holes. Ideally, the shield should be a spherical shroud enveloping the source or victim and bonded to ground by soldering or welding. More practically, the shield is configured as a rectangular box, cylinder, or sheet barrier. It may be solid, screen, braid, metal foil, metallic tape, impregnated plastic, or even be coated with a conductive paint or spray. It is usually installed with rivets or screws atop an RF gasketing material. The important thing to keep in mind is that to achieve optimum shielding for both E-field and H-field interference, the maximum degree of continuous conductivity to ground must be ensured.

Table 4-7. K Correction Factors for Iron and Copper in dB

	Shield Thickness (Mils)	*Frequency*					
		60 Hz	*100 Hz*	*1 kHz*	*10 kHz*	*100 kHz*	*1 MHz*
Magnetic fields copper ($\mu = 1$, $\sigma = 1$)	1	−22.22	−24.31	−28.23	−19.61	−10.34	− 2.61
	5	−21.30	−22.07	−15.83	− 6.98	− 0.55	+ 0.14
	10	−19.23	−18.59	−10.37	− 2.62	+ 0.57	—
	20	−15.85	−13.77	− 5.41	+ 0.13	− 0.10	—
	30	−12.55	−10.76	− 2.94	+ 0.58	—	—
	50	− 8.88	− 7.07	− 0.58	—	—	—
	100	− 4.24	− 2.74	+ 0.50	—	—	—
	200	− 0.76	+ 0.05	—	—	—	—
	300	+ 0.32	+ 0.53	—	—	—	—
Electric fields and plane waves copper ($\sigma = 1$, $\mu = 1$)	1	−41.52	−39.31	−29.38	−19.61	−10.33	− 2.61
	5	−27.64	−26.46	−15.82	− 6.96	− 0.55	+ 0.14
	10	−21.75	−19.61	−10.33	− 2.61	+ 0.57	—
	20	−15.99	−13.92	− 5.37	+ 0.14	− 0.10	—
	30	−12.73	−10.73	− 2.90	+ 0.58	—	—
	50	− 8.81	− 6.96	− 0.55	+ 0.14	—	—
	100	− 4.08	− 2.61	+ 0.51	—	—	—
	200	− 0.62	+ 0.14	—	—	—	—
	300	+ 0.41	+ 0.58	—	—	—	—
Magnetic fields iron ($\mu = 1000$, $\sigma = 0.17$)	1	+ 0.95	+ 1.23	− 1.60	− 1.83	—	—
	5	+ 0.93	+ 0.89	− 0.59	—	—	—
	10	+ 0.78	+ 0.48	+ 0.06	—	—	—
	20	+ 0.35	+ 0.08	—	—	—	—
	30	+0.06	− 0.06	—	—	—	—
	50	—	—	—	—	—	—
Electric fields and plane waves iron ($\mu = 1000$, $\sigma = 0.17$)	1	−19.53	−17.41	− 8.35	− 1.31	—	—
	5	− 6.90	− 5.17	+ 0.20	—	—	—
	10	− 2.56	− 1.31	+ 0.36	—	—	—
	20	+ 0.16	+ 0.54	—	—	—	—
	30	+ 0.58	+ 0.42	—	—	—	—
	50	+ 0.13	—	—	—	—	—

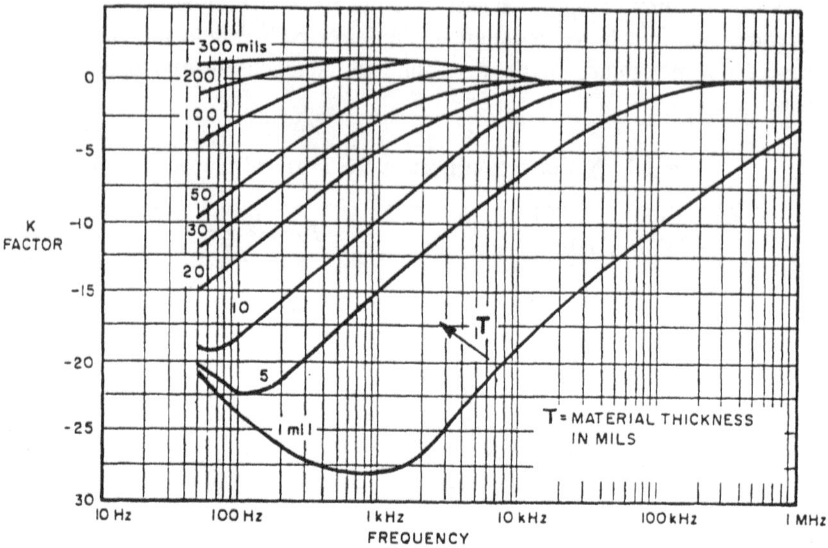

Figure 4-9 K Correction Factor for Copper Magnetic Field

4-2.1.2 Shielding Methods and Materials

a. *Multiple Layer Metallic Shields*—Low frequency H-field EMI shielding frequently requires specialized materials and techniques. While high permeability alloys such as Mumetal do provide good absorption loss attenuation for the weaker low frequency interferences, strong magnetic energy sources necessitate a combination of reflection and absorption losses to achieve sufficient shielding. Since single-wall reflection losses are quite small at low frequencies, the solution is to present multiple barriers to the interfering signal, thereby promoting a succession of reflection losses resulting from each boundary (along with accumulative absorption losses). The composite partition thus offers a superior shielding effectiveness option to the single-wall shield of greater overall thickness. Moreover, since ferromagnetic metals tend to saturate at a maximum flux level, the interspersing of ferromagnetic and nonferromagnetic layers offers better attenuation characteristics over an individual saturated metal. Generally, metals such as copper, iron, Mumetal, and CoNetic are clad together (but separated by air space or solid dielectric material) to form a highly effective electromagnetic shield affording 100 dB or more of interference reduction.

Low frequency H-field interference is produced aboard ship by such items as motors, transformers, solenoids, and coil inductors. One practical method used to verify the source location and suppress this EMI at its origin is to shroud a suspected interference generator with an easily formed metallic shield. Various types of flexible sheet and foil materials are available which can be cut, shaped, and taped in place to determine experimentally the thickness required, the number of layers, and the best metal or alloy needed for effective shielding.[10]

b. *Perforated Metallic Shielding*—Because of requirements such as ventilation and visual monitoring, there are some instances aboard ship where solid metal EMI shields cannot be used; for example, where equipment cooling calls for continuous air circulation through partitions, and where instrument meters and display scopes must be viewed through transparent covers. Satisfactory shielding, though never as good as that of solid barriers, can be achieved using wire mesh screens across ventilation openings. Likewise, fine woven mesh may be implanted in, or laminated with, glass to provide a transparent medium across viewing apertures.

Normally the shielding effectiveness of a metal screen decreases as frequency increases, and with increasing hole size (wire spacing). Magnetic energy, however, is an exception: the shielding effectiveness *increases* with increasing frequency for H-fields (i.e., the attenuation is small at low frequencies), and, of course, increases with the permeability of the metal in use. In either event, calculating shielding effectiveness for metal screens is so tedious that it is more practical to select material readily available from vendors proposed for meeting the attenuation requirements of a particular situation. Then, after applying the screen, the resultant field strength is measured and compared to the EMI strength prior to screen installation. In this manner the shielding adequacy of the metal screen can be easily verified.

In some situations ventilation and cooling requirements necessitate an unimpeded, high air flow with screen holes so large that adequate shielding effectiveness cannot be met with wire mesh. For these cases metal honeycomb shields are recommended. Honeycomb barriers make use of waveguide transmission line theory to determine hole size and depth so that the shield openings act to attenuate greatly a wide band of potential EMI below the waveguide cutoff frequency.

The effectiveness of the honeycomb is a function of the hole size, depth, spacing (i.e., hole number), and type metal. Honeycomb panels are heavier and more expensive than wire mesh screens; they offer better shielding effectiveness, however, and greater structural strength. They are particularly effective at the higher frequencies. For example, at 10 GHz, a ¼-inch diameter honeycomb tube 1 inch in depth (thickness) provides 102 dB attenuation, and, at the same frequency, a ½-inch diameter tube (higher air circulation) 2¼ inches in depth still gives 100 dB attenuation. Moreover, even at the lower frequencies, honeycomb panels have fairly good shielding effectiveness, as seen in Table 4-8 for a steel honeycomb screen with hexagonal openings ⅛-inch wide and ½-inch deep.

Table 4-8. Shielding Effectiveness of Steel Hexagonal Honeycomb ⅛-Inch Openings ½-Inch Thick

Frequency (MHz)	*Shielding Effectiveness (dB)*
0.1	45
50	51
100	57
400	56
2,200	47

c. *Metallized Surfaces Shielding*—An interesting change has occurred in the packaging of electronic circuitry during the final quarter of the twentieth century. The change, so widespread and rapid as to be perhaps unnoticed by the casual observer, is the virtual worldwide use of molded plastic enclosures. In the past it was common practice, particularly for high-quality military equipment, to assemble electronic components on a metal chassis and then house the completed package within a heavy metal cabinet. The metal case afforded a good degree of electromagnetic shielding as well as excellent structural ruggedness. Recently, however, lightweight molded plastics have been universally adopted for encasing the new generations of solid-state microelectronic devices. This trend is readily apparent in the proliferation of such modern office items as word processors, desktop computers, printers, and various peripheral equipment. To a lesser but significant extent, even some contemporary large shipboard electromagnetic systems are now being enclosed in plastic cabinets.

In addition to being very light in weight, plastic cases are relatively inexpensive, tough, and attractively contoured and colored. Yet they are notably deficient in one important aspect: they are transparent to RF energy.

That is, unless carefully prepared, they offer no EMI shielding. Therefore, special measures must be taken to convert the plastic shell into a continuously conductive envelope so as to protect otherwise vulnerable internal electronic circuitry from external electromagnetic disturbances. Likewise, internally generated interference must be prevented from escaping into the environment.

Numerous engineering techniques have been developed to transform plastic enclosures into effective EMI shields. It must be assumed that one or more of these techniques would be incorporated in equipment specified, designed, and procured for naval shipboard use. Nevertheless, there has been at least one important incident where an electronic warfare system housed in a plastic enclosure emitted broadband noise from the rear of the case sufficient to cause severe EMI to other topside combat systems. Retrofit shielding was necessary to quell the interference. Similarly, corrective maintenance and repairs to other shipboard equipment is required from time to time.

One method popularly employed to provide good enclosure shielding is to mix metal fibers such as aluminum, stainless steel, carbon, or graphite in with the liquid plastic during the molding process to create an electrically conductive composite material. Because their high length-to-diameter ratio makes them more efficient conductors, metallic fibers are preferred over other metal fillers such as powder, granules, or flakes.

If the plastic has no embedded metal for shielding, several types of metallic coating techniques are available for plastic surfaces, including vacuum metallizing, wire-arc and flame metal spraying, electroless metal plating, and the use of metal-foil linings, metal-coated fabrics, and conductive paints. The most commonly used metals in these applications are aluminum, copper, silver, zinc, nickel, iron, carbon, carbon steel, stainless steel, nickel steel, and graphite.

1. *Metal Foils and Metallized Tapes*—Various types and thicknesses of metal foil are commercially available and are easily formed and applied to furnish satisfactory shielding. The most effective foil shields are flexible laminates, in which such foils as copper or aluminum are sandwiched between reinforcing plastics or paper films. Accordingly, the foils provide good EMI attenuation, and the dielectric film substrates act as electric insulating materials.

   For small items, or for quick stop-gap repairs, metallized foil tape with pressure-sensitive adhesive may be handily used, as it easily conforms to nearly any shape of object in need of shielding.

2. *Fiber-Coated Fabrics*—An alternative to foil, also lightweight, flexible, and simply applied, is metallized fabric. Fibrous material such as rayon, cotton, polyester, polyacrylate, polyamide, polyurethane, or even glass or carbon is used as a base and plated with a microthin metal membrane such as gold, silver, copper, nickel, cobalt, or chrome.[11] By proper combination of fiber, metal coating, and coating thickness, the desired EMI protection is achieved, with a shielding effectiveness typically 50 dB over a frequency range from 100 kHz to 1 GHz.

   The selected fabric can be purchased with an adhesive backing and quickly cut to conform to any practicable size or shape. Moreover, since EMI fabrics are lightweight, breathable, and washable, they may be employed as screen draperies and wall coverings and even may be fashioned into outerwear for protection against EM radiation hazards.
3. *Vacuum Metallizing*—As the name suggests, this method of coating is achieved through evaporation in a high vacuum to deposit a uniform metallic film on a nonconductive surface.[12] Before metallizing, the object to be shielded is treated with a chemical base coat and baked to allow good adhesion of the metal substance. Then, with aluminum as the most commonly used evaporant, the vacuum process is carried out to coat the enclosure surface.

   Closely related to vacuum metallizing is a technique called sputtering. It, too, is accomplished in a high vacuum, but, rather than using the evaporation process, a metal target is bombarded by electrically excited argon ions that dislodge and sputter the metal atoms to deposit them on the desired nonconductive surface.[13]

   Generally a chrome film base is applied initially to the bare plastic surface, followed by a copper alloy for high conductivity, and, finally, by another coating of chrome for corrosion resistance. In this way a combined medium of good conductivity with oxidation protection provides highly effective EMI shielding.
4. *Wire-Arc and Flame Spray Shielding*—In the wire-arc spray method, wires (zinc wires are commonly used) are kept electrically isolated and continuously fed into an operating gun so that only the wire ends maintain contact. At a precise point the ends are melted instantaneously by an intense arc, and the molten, atomized metal is directed to the target by a high pressure jet of air.[14] Upon contacting the plastic casing, the molten zinc quickly solidifies to form a dense, conductive, metallic coating.

   A somewhat related technique known as flame spraying differs in that it utilizes an acetylene flame rather than an electric arc. The flame method has the disadvantage of heating up the plastic surface, and extreme caution must be exercised to prevent distortion and warping.

5. *Electroless Plating*—A chemical process often used to coat nonconductive surfaces is called electroless plating. Because the article to be coated is dipped in an aqueous solution, this method offers the immediate advantage of depositing a metallic film on both the inner and the outer surfaces simultaneously in a single application. The shielding effectiveness is thereby much improved over single-shielded surfaces. Electroless plating differs radically from electroplating, where an external electrical source (e.g., a dc rectifier supply) is used to plate metal upon a *conductive* surface. Electroplating cannot be used on nonconductive materials. Furthermore, electroless plating is not to be confused with so-called immersion plating, which involves replacement of one metal of higher electromotive potential by another of lower electromotive potential.[15]

   In electroless plating the aqueous solutions are a mixture of reducing agents and a reducible metal, and the chemical reaction occurs only at the formerly nonconductive surface. The item to be coated is first prepared by an etching process and is neutralized. Then a catalyst is employed to activate and deposit an electroless metal such as copper or nickel alloy. The thickness of the electroless coating is determined by the length of time the object remains immersed in the plating bath.

   As a result of its simplicity, electroless plating will coat nearly any type of enclosure configuration, no matter the complexity, with a uniform metallic film. If additional shielding is desired, electroplating may be applied on top of the electroless metal film. For example, a layer of electroless copper can give the needed high conductivity while an overcoat of electroplated nickel alloy allows good corrosion resistance, so that the multilayer metallic combination affords excellent overall shielding effectiveness.

6. *Conductive Paint Coatings*—Another technique for metallizing plastic enclosures is the use of sprays containing conductive metal particles. In this procedure the metal coat is normally applied to the interior walls of the cabinet. Several conductive substances are available[16] for use in such sprays as acrylic lacquer for fast-drying good adherence to most thermoplastics, or in urethane paints where greater coating hardness is required.

   (a) *Graphite*—Mixed in paint solutions such as acrylic, polyurethane, epoxy resins, or flexible rubber compounds, graphite yields a shielding effectiveness of 30 to 40 dB over a range of 200 MHz to 1 GHz, and 40 to 60 dB for frequencies above 1 GHz.

(b) *Copper*—Copper has high conductivity, nearly as good as silver, but it oxidizes too easily unless specially treated. It has excellent shielding characteristics at the lower frequencies: 87 dB at 30 MHz, 84 dB at 300 MHz, and 36 dB at about 1 GHz.

(c) *Nickel*—Where oxidation and corrosion must be avoided, nickel is frequently selected. Because of their high permeability, and, therefore, high H-field absorption, nickel compounds are currently the predominant choice of spray paint coatings. Shielding effectiveness for nickel sprays is 40 dB or greater from 20 MHz to 1 GHz.

(d) *Silver*—Silver, of course, has the disadvantage of being relatively expensive as a conductive ingredient. It offers, however, two distinctive merits: it is highly conductive and it resists oxidation. Consequently it is an excellent shielding element. Recent developments have reduced the cost of using silver by taking advantage of RF skin-effect phenomena; that is, in realizing that higher frequency currents are concentrated on the outer surface of metal conductors. Accordingly, microspheres of hollow glass or ceramic are transformed into highly efficient electromagnetic conductors by a thin coat of silver.[17] With a diameter of 50 microns and silver coat thickness of one microinch, the miniature spheres mix easily in the paint solutions and facilitate spraying with conventional equipment.

Silver-coated spheres are highly effective electromagnetic shields. EMI energy penetrating a sphere is dissipated by destructive interference of the multiple internal reflections. And overall surface conductivity of the plastic enclosure results from the microspheres' being in natural contact to form conductive chain networks. If a denser coating is preferred for optimum shielding effectiveness, the hollow spheres may be filled with a magnetic material to ensure tight packing by magnetic attraction. Silver paints, with an effectiveness of over 60 dB across the frequency spectrum, are often selected for military equipment as offering the best shielding quality.

For quick field repairs and experimental test purposes aerosol spray conductive paints are also available. But a significant disadvantage of spray paint coatings is that, compared to the other forms discussed above, they are easily scratched and marred to disrupt shielding integrity.

d. *RF Gasket Shielding*—Figure 4-10 is a popularly used illustration showing several apertures that are likely to cause shielding degradation of an enclosure. If the rectangular outline is viewed as an equipment cabinet or a laboratory screenroom or, ultimately, a ship hull, it is apparent that, no matter how high the degree of wall shielding, a number of potential weak points exist. Depending upon the enclosure's purpose, there will be powerline and cable penetrations, antenna connections, ventilation openings, instrument view-ports, and entryways. Each access represents a deterioration of shielding integrity. Viewports and ventilation openings may be shielded by mesh or honeycomb screens; control shafts can use waveguide below cutoff attenuators as discussed in Paragraph b of this section; and cables and powerlines use shielding and filtering methods, to be addressed later in this chapter. Of interest to us now are the means whereby edges, cracks, and seams are shielded by use of RF gasketing.

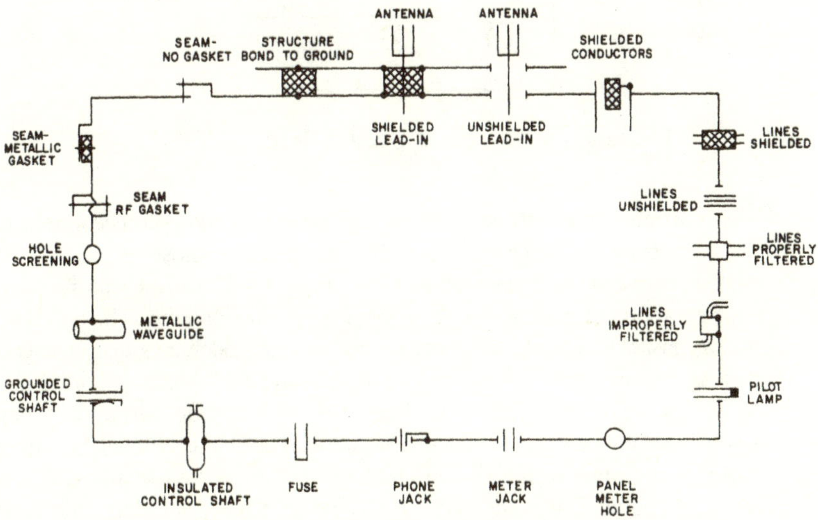

Figure 4-10 Typical Shielded Enclosure Discontinuities

In some instances seams can be protected from EMI penetration without employing RF gaskets by proper mating of the metal surfaces. Special care must be taken in such cases to establish a good bond between the metal surfaces. This must be achieved by avoiding permanent contact of dissimilar metals (to preclude galvanic corrosion); by ensuring that the bare metal surfaces to be connected are thoroughly clean and dry; and by maintaining sufficient fastening pressure with screws, rivets, soldering, or continuous welds. A preferred technique is depicted in Figure 4-11; e.g., folding and overlapping the edges of thin walled enclosures, followed by soldering or welding.[18] The resultant seam then has a shielding thickness triple that of the enclosure wall.

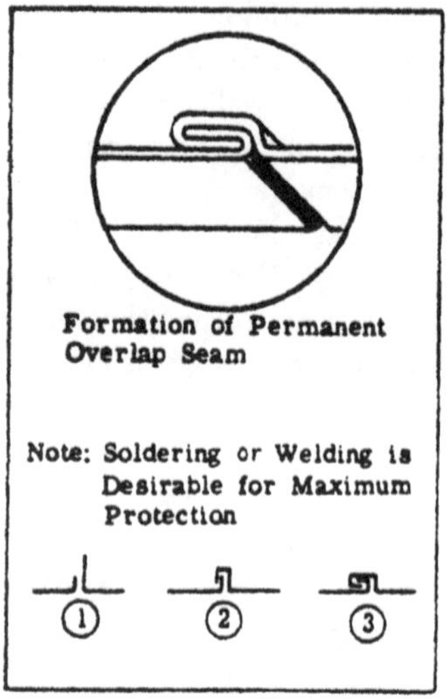

Figure 4-11 Formation of Permanent Overlap Seam

The most effective method for ensuring that seams and entrance cracks do not unacceptably reduce shielding effectiveness of enclosures is to use RF gaskets between the connected surfaces. Figure 4-12 shows representative configurations of gasketed surfaces commonly encountered. The two types of RF gasket materials most often used are knitted wire mesh and conductive rubber stock.[19] Knitted wire mesh is available in phosphor bronze, Monel (a nickel-copper alloy), aluminum, tinned copper-iron alloy, and silver-plated brass. The conductive elastomers use silver, nickel, or carbon particles as fillers in the silicone rubbers. Various cross-sections of wire mesh or rubber elastomer gaskets may be selected, including round, elliptical, rectangular, P-shape, U-channel, and tubular.

Whenever there is the slightest chance that the enclosure might be exposed to dampness, such as in shipboard topsides, it is imperative that the RF gasket also be a moisture-tight seal. Electrolytic conditions must be prevented at the flanges or seams. In this case the conductive elastomer gaskets are better sealing materials than the wire mesh types, although combination solid rubber and wire mesh gaskets can be very effective for some configurations. For the extreme salt spray shipboard environments, dual or combination gaskets should be used. Care must be taken in the selection of gasket metal type in these environments. For example, tin-plated, copper-clad steel mesh should be avoided, because if the copper becomes exposed due to abrasion, the seam will rapidly oxidize. Tin-plated, copper-clad steel is, however, an excellent gasket for attenuation of high H-fields where salt spray conditions are not present. The best gasket material for

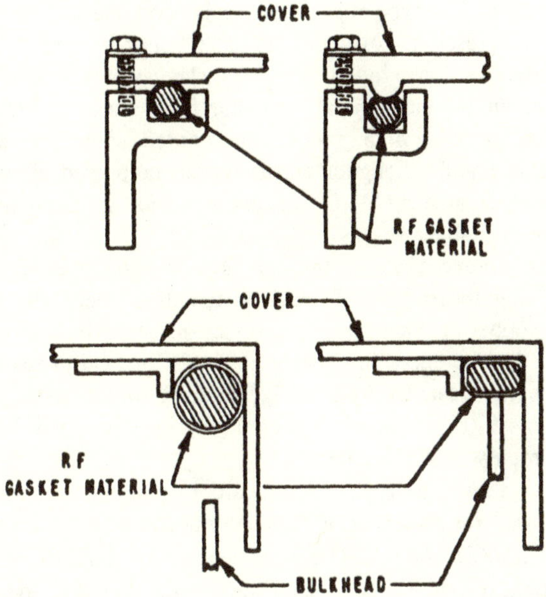

Figure 4-12 Gasketed Surfaces (from USAF EMC Design Handbook DH 1-4)

shipboard corrosion resistance at aluminum flanges is a silver-plated, aluminum-filled elastomer.

Because aluminum is so widely used in ship systems design, it merits special note. When installed between aluminum surfaces, virtually all types of EMI gaskets have a high potential for creating galvanic corrosion at the flanges and seams.[20] Gaskets made only of aluminum are usually avoided because aluminum generates an impenetrably hard, nonconductive oxide. Yet, in most other choices where the gasket is made of metals dissimilar to aluminum (e.g., Monel, tin, carbon, silver, or beryllium copper), the dissimilarity readily promotes galvanic cell corrosion. The result is increased resistance between the shielding gasket and the enclosure surface and decreased shielding effectiveness. For this reason silver-plated aluminum conductive rubber gaskets are recommended for aluminum seams.

It is critical for good enclosure gasketing that adequate and even pressure be applied between the shielded surfaces.[21] This is accomplished by providing a rigid surface at the contact point, using several evenly spaced fasteners, and incorporating self-retaining grooves to hold the gasket in place. In some difficult situations the use of adhesives or spot welding is recommended. The type of gasket cross-section selected is also a function of proper compression. For pressures of 20 psi or less, knitted wire mesh of round cross-section, knitted wire mesh sleeves over sponge elastomers, or solid conductive elastomers in tubular form are best. When the seam compression must be high, rectangular knitted wire mesh and solid P-shape, U-shape, and rectangular elastomers are employed. If over-compression is an anticipated problem, gaskets with built-in metal stops of brass or steel are available.

Where frequent access through an enclosure opening is required, flexible spring finger stock should be installed. Generally available in beryllium copper or other highly corrosion-resistant alloys, finger gaskets are designed to be exceptionally durable to withstand thousands of operations. They are fastened by adhesive, epoxy, solder, rivets, or clips.

Table 4-9 (from [18]) lists the major advantages and deficiencies of some commonly used RF gaskets.

Figure 4-13 illustrates typical shipboard installations of conductive RF gasketing used for shielding of flange connections.

e. *Wire and Cable Shielding*—The large quantity of complex, sensitive electronic systems installed aboard naval warships gives rise to an extraordinary number of cable and wiring interconnections. To carry out its necessary functions, shipboard electronic equipment requires cabling and wiring for transmission of electrical information such as audio, video, pulse, and control signals; operating power; and RF radiation and receive energy. To do so, cableways must traverse virtually every below-deck compartment (see Figure 4-14) and, in many cases, must penetrate the main deck to run from stem to stern, to and from a variety of topside items and spaces, reaching even to the very tips of masts (see Figure 4-15), where they may be exposed to enormously intense electromagnetic fields.

   Unfortunately, in being so ubiquitous, cables and signal wires are also expedient bearers of bad news. That is, because of their own electromagnetic nature, they act as convenient antennas and transmission lines to intercept and transport radiated and conducted EMI straight into the weakest spots of electronic equipment. This tendency to aid and abet system degradation results in cables and wires being given special attention. In shipboard systems engineering design and installation, cable shielding and routing practices are of extreme importance.

   1. *Cable Types and Terminations*—Most below-deck interference is generated by shipboard power systems and other frequency sources below 100 kHz, while in the topside, radiated EMI occurs generally at HF and above. Therefore, because of the high probability of encountering

several forms of EMI ranging in frequency from powerline to microwave, and because of their natural susceptibility, shielded cables are invariably used. The principal types available are shielded single wire, shielded multiconductors, shielded twisted pair, and coaxial. Several variations of these are used to protect against both magnetic fields and stray RF. Coaxial, even though meant primarily for high frequency transmission, is an excellent low-frequency shielded cable and is frequently selected because of its adaptability and comparatively low cost. If subjected to very strong levels of EMI, however, even coaxial might not afford adequate shielding. In such cases, more complex cable designs are employed; e.g., twinaxial, triaxial, and quadraxial.

Table 4-9. Characteristics of Conductive Gasketing Materials

Material	*Chief Advantages*	*Chief Limitations*
Knitted wire mesh	Most resilient all-metal gasket (low flange pressure required). Most points of contact. Available in variety of thicknesses and resiliencies, and in combination with neoprene and silicone.	Not available in sheet (certain intricate shapes difficult to make). Must be 0.040 inch or thicker. Subject to compression set.
Brass or beryllium copper with punctured holes	Best of corrosion protection films.	Not truly resilient or generally reusable.
Oriented wires in rubber silicone	Combines flexibility and RF seal. Can be effective against corrosion films.	Might require wider or thicker size gasket for same effectiveness. Effectiveness decreases with mechanical use.

Table 4.9 (cont'd)

Material	*Chief Advantages*	*Chief Limitations*
Aluminum screen impregnated with neoprene	Combines flexibility and conductive seal. Thinnest gasket. Can be cut to intricate shapes.	Very low resiliency (high flange pressure required).
Soft metals	Least expensive in small sizes.	Cold flows, low resiliency.
Metal foil over rubber	Has advantage of the resiliency of rubber.	Foil cracks or shifts position. Generally low absorption loss yielding poor RF properties.
Conductive rubber (carbon filled)	Combines flexibility and conductive seal.	Provides moderate absorption loss.
Conductive rubber (silver filled)	Combines flexibility and RF seal. Excellent resilience with low compression set. Reusable. Available in any shape or cross section.	Not as effective as metal in magnetic fields. May require salt spray environmental protection.
Contact fingers	Best suited for sliding contact.	Easily damaged. Few points of contact.

LIST OF MATERIAL

ITEM NO.	PART	SPECIFICATION	NOTE
1	GASKET MATERIAL, CONDUCTIVE	COMMERCIAL	1

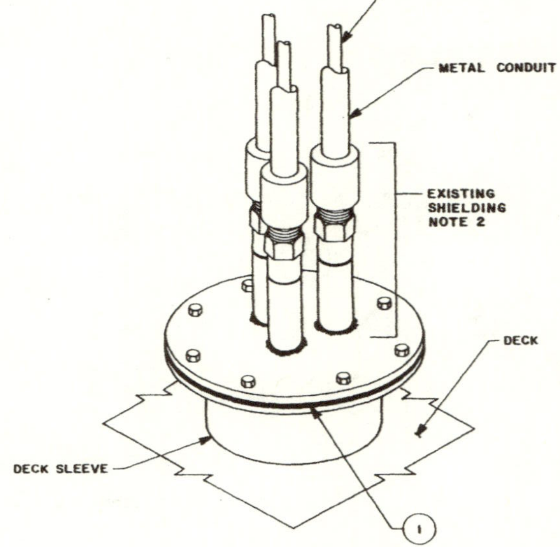

POWER CABLE DECK PENETRATION

NOTES:

1. WHEN DECK SLEEVE IS USED INSTALL CONDUCTIVE GASKET MATERIAL IN ACCORDANCE WITH SHEET 1, NOTE 5.
2. EXISTING CONFIGURATIONS MAY DIFFER AMONG INDIVIDUAL SHIPS IN THE CLASS. DO NOT REWORK EXISTING CONFIGURATIONS TO MATCH THROUGHOUT THE CLASS, BUT INSURE THAT ALL CABLES ARE PROPERLY SHIELDED.
3. WEATHER SEAL ALL GASKET EDGES WITH "PERMATEX II", "SCOTCHCLAD" OR "BMS 595".

NAVSEA DRAWING NO.	REV	SHEET
806-6215591	A	14

Figure 4-13(a) Shipboard Installation of RF Gasketing at Flanges

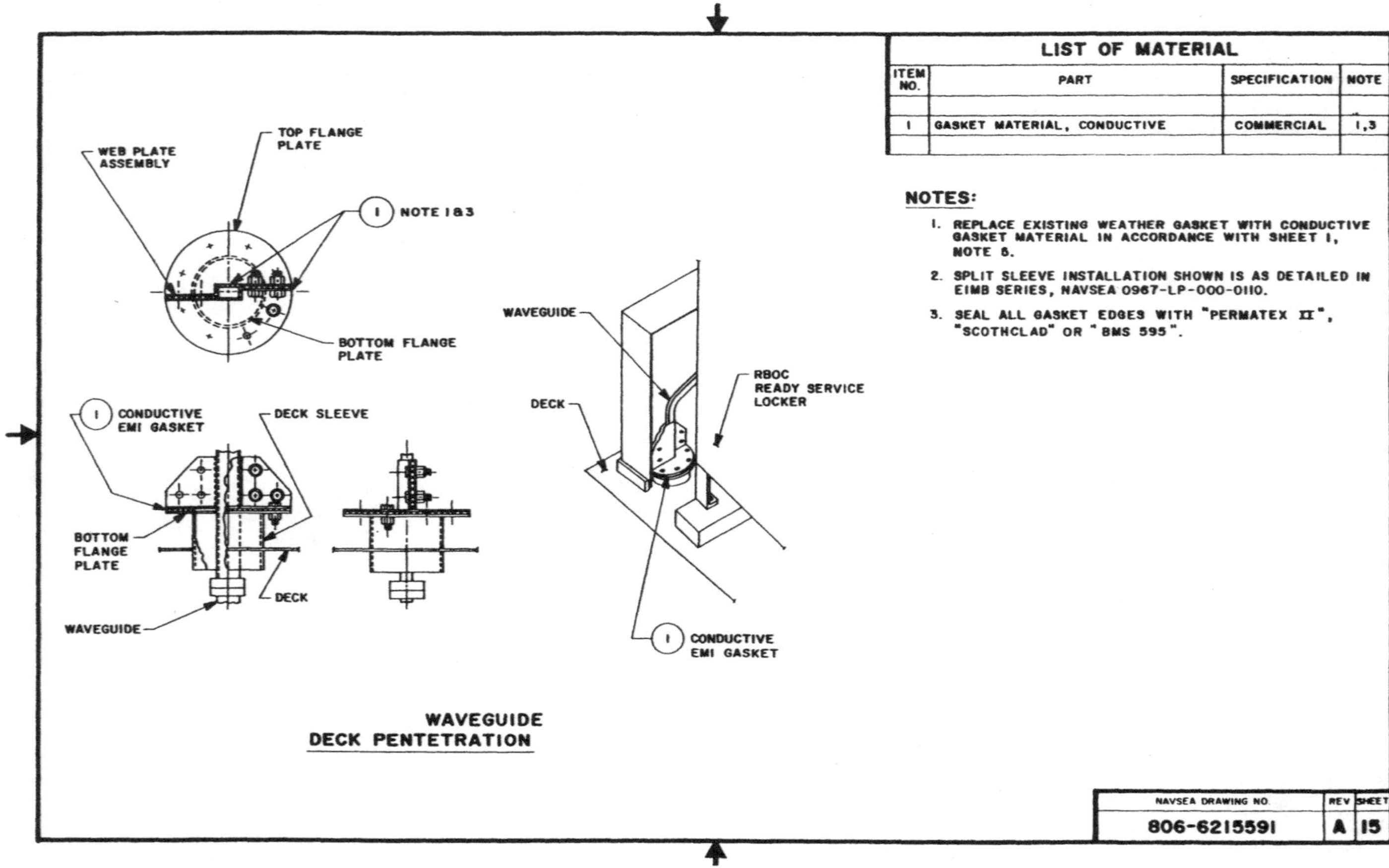

Figure 4-13(b) Shipboard Installation of RF Gasketing at Flanges

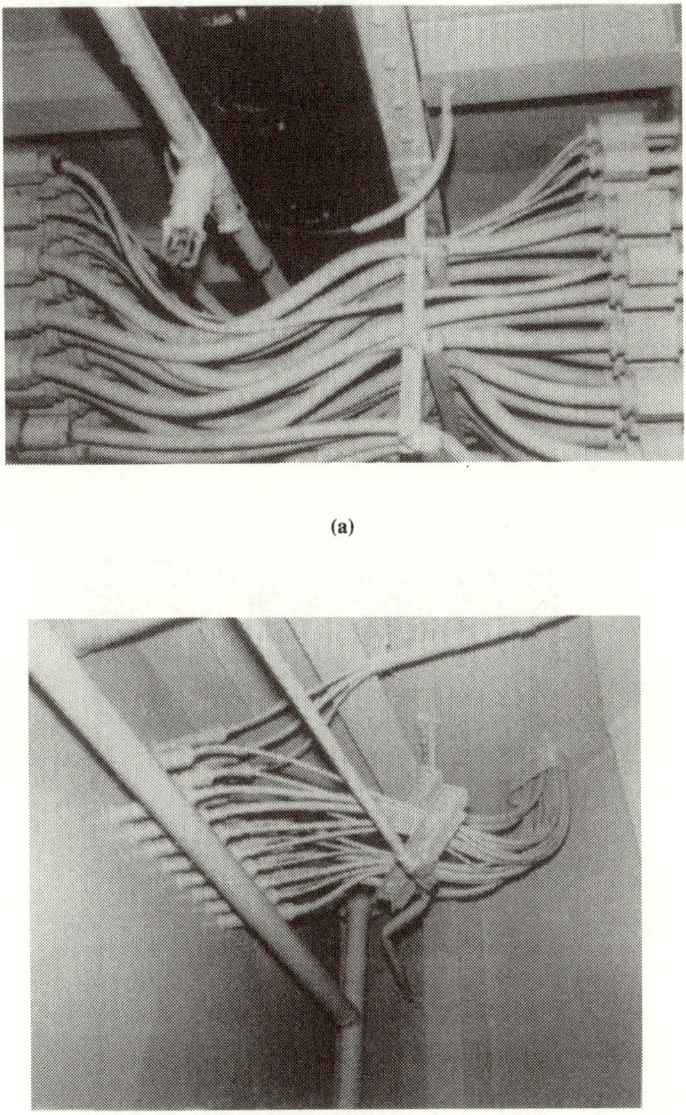

(a)

(b)

Figure 4-14 Cables Traversing Below-Deck Compartments

(a)

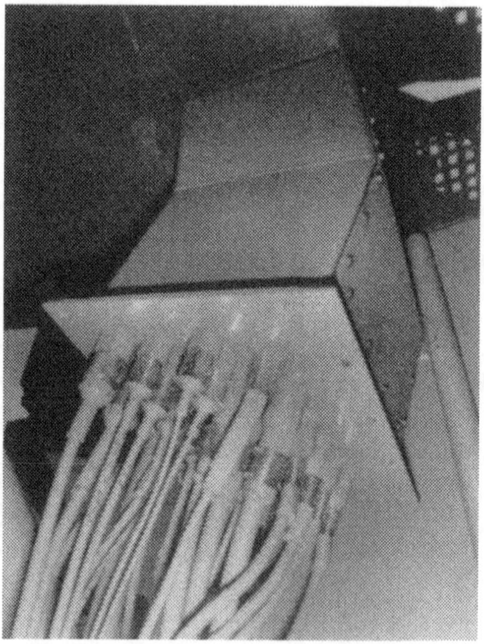

(b)

Figure 4-15 Cables Entering and Running Up Ship Mast

Triaxial is simply a coaxial cable with an added outer copper-braid sleeve that encloses the inner coaxial conductors to allow increased shielding. The outer sleeve is grounded so as to shunt coupled noise energy and ground loop currents away from the inner signal-carrying conductors. As a result, the signal-to-noise ratio is substantially improved over that of ordinary coaxial.

Twinaxial is a double-conductor, twisted, balanced wire cable also having an outer metallic braid to shield the enveloped wires. Because the twist configuration causes cancellation of induced noise energy, good protection against penetrating H-field interference is provided. Unfortunately, high frequency transmission losses limit the use of twinaxial to below about 15 MHz.

Where severe low-frequency EMI is anticipated, another shielded cable option known as quadraxial is effective in many cases. Quadraxial is a configuration in which twinaxial wiring is enclosed within a second outer shield. In this mode, the outermost braid is connected to the "earth ground," and the inner shield is connected to the electronic system ground to yield overall high protection from interference.

The effectiveness of shielded cable cannot be fully realized, however, unless the cable is carefully and correctly terminated. Sloppy dressing of the conductors, poor grounding of the braided shield, and otherwise improper termination of the cable can degrade the shielding effectiveness by as much as 30 dB. Indeed, unsuspected causes of shielding degradation can often be traced directly to incorrect RF grounding procedures. In an otherwise satisfactorily shielded system, RF currents are conducted along the cable shields and coupled directly into the system equipment from inadequately terminated connectors. This is especially evident at high frequencies where it is imperative that multiple grounding be employed to minimize RF currents along the shield. To terminate a cable properly, the entire periphery of the braided shield must be grounded to a low-impedance reference. This will ensure reduction of RF currents to a minimum at the termination surface. Soldering of the braid is discouraged in favor of using crimping rings or tapered cone compression methods as shown in Figures 4-16 and 4-17. Backshells are commonly used for added protection of the termination, but it is important that the metal composition of the backshell match that of the cable connector and conductors to preclude galvanic corrosion and shielding deterioration due to mating of dissimilar metals.

2. *Cable Identification and Spacing*—Second only to proper selection, termination, and grounding of shielded cable is the importance of careful spacing and routing of shipboard cables. In times past it may have been that the bundling and routing of wires and cables throughout a ship was

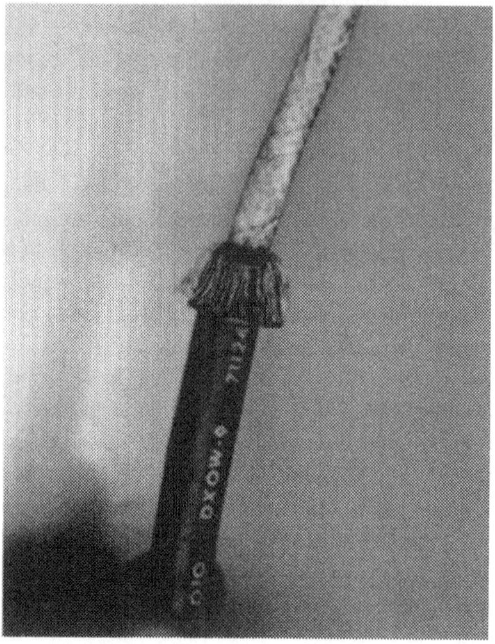

Figure 4-16 Proper Dress of Metallic Shield Over Tapered Compression Cone

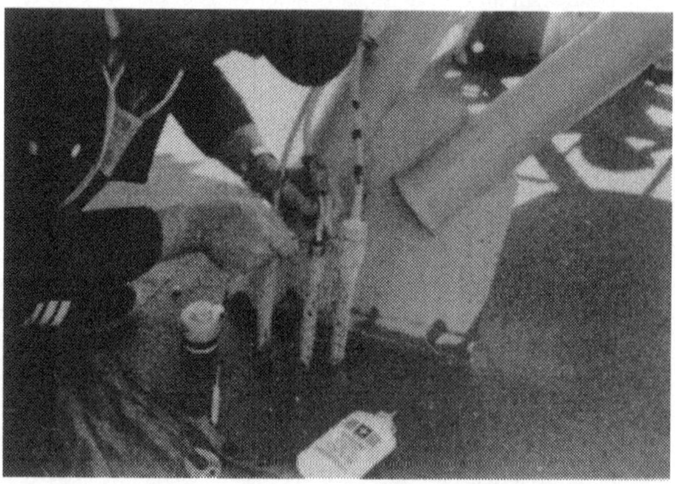

Figure 4-17 Careful Grounding and Sealing of Cable at Deck Entry Points

simply a matter of reaching one point from another by the shortest, least complicated route. Not so in our modern warships. Cables are appreciated now as an integral part of the ship's electronic subsystems. Justly so, the practice of shipboard cable routing involves identification, separation, and proper placement. The best way to minimize coupling between cables is to separate the likely radiators as widely as possible from the potential susceptors and, ideally, to orient them at 90° to each other.

Initially, shipboard cables must be segregated, identified, and tagged as to type and function. The Navy has categorized cables by whether they represent probable radiators of interference (R-types), or are potential susceptors of EMI (S-types). The cables are then designated according to use as specified in Tables 4-10 and 4-11. This information is subsequently used to determine the necessary cable-to-cable spacing to preclude EMI coupling (in accordance with the methods outlined in detail in [22]).

Actual naval shipboard incidences have been documented, for example, where cables carrying HF frequencies have coupled RF energy into power wiring, and cables transmitting radar modulator pulses have induced interfering signals in radio control lines to degrade communications significantly. Adequate cable-to-cable spacing had been neglected.

3. *Cable Conduit Shielding*—Naval engineering practice in general is to avoid the use of any external shielding devices if at all possible. Preferably, sufficient shielding between R- and S-type cables should be achieved by judicious routing and spacing so that the need for additional shielding is minimized.

   Nevertheless, there are instances where satisfactory shielding protection cannot be obtained by either the cable's inherent shielding material or by separation between cables. This is particularly the case where cables are exposed to intense RF fields and where an increase in skin depth conduction at low frequencies necessitates the use of thicker shielding. For such events there is no alternative except to enshroud the cables within metal conduits, a last resort because conduits add much weight and cost. (Note that the use of so-called armored cable, i.e., cable jacketed by a flexible, hard, braided metal outer shell, has been discontinued in shipboard installations because it is a severe source of intermodulation interference and broadband (RF arcing) noise.)

   The two types of pipe conduit specified for shipboard application are rigid and flexible. As seen in Figure 4-18, these conduits are highly effective shields even at power frequencies, especially above 10 kHz.

Table 4-10. R-Type Cable Categories

Cable Category	Category Description
R1	Shipboard cables that carry 60-Hz power.
R2	Shipboard cables that carry 400-Hz power.
R3	All transmitting systems operating below 100 kHz.
R4	Transmitting systems and triggering circuits operating above 100 kHz and using RG-type coaxial cables.
R5	Cables used to carry audio signals whose maximum values exceed 0.1 volt. Typical components are announcing circuits, ac recorders, loudspeakers, call bells, and alarm bells.
R6	Cables that carry 60-Hz synchro signals, 60-Hz control signals up to 0.5 amp, and 60-Hz indicator signals. NOTE: Any 60-Hz control signal over 0.5 amp must be classified in the R1 category.
R7	Cables that carry 400-Hz synchro signals, 400-Hz control signals up to 0.5 amp, and 400-Hz indicator signals. NOTE: Any 400-Hz control signal over 0.5 amp must be classified in the R2 category.
R8	Cables used to carry digital data.
R9	Cables that carry dc.

Table 4-11. S-Type Cable Categories

Cable Category	Category Description
S1	Receiving systems operating in the frequency band 10 kHz to 100 kHz.
S2	Same as S1 except different type of cable.
S3	Receiving and video systems operating above 100 kHz.
S4	Receiving systems operating below 10 kHz.

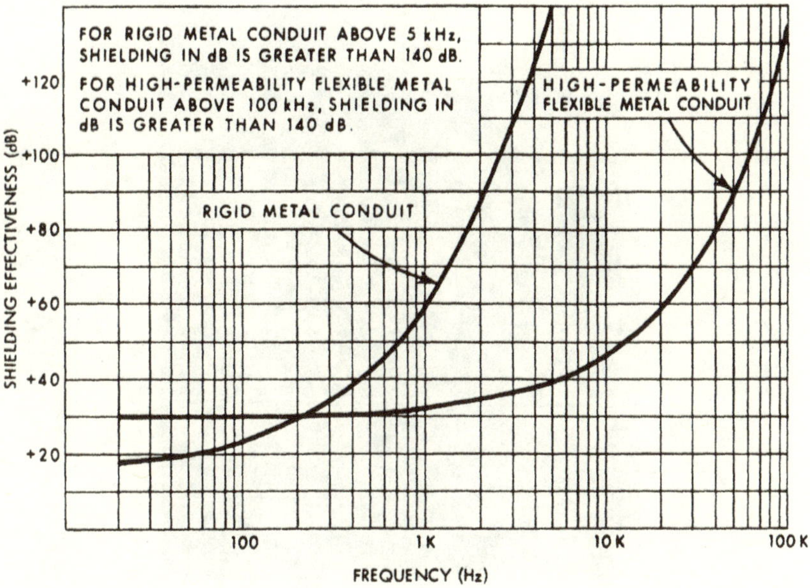

Figure 4-18 Conduit Shielding Effectiveness

Navy specifications require that rigid conduit used for EMI shielding be seamless steel pipe in accordance with MIL-T-20157 and have a wall thickness of not less than 0.120 inch.[23] The conduit must be made shock- and vibration-resistant by the use of rubber padded pipe hanger cushions and be grounded by means of bond straps. To absorb further shock and vibration, rigid conduit should be terminated with about 30 inches of flexible conduit at the entry to equipment enclosures, bulkhead stuffing tubes, and hull fittings.

Flexible conduits are used at frequencies below 100 kHz, where low-level signal cables must be well shielded from strong magnetic fields. Type-1 nonjacketed flexible conduit is recommended where the cable being shielded is not susceptible to EMI from currents flowing along the pipe, and where stray currents are minimal. But for extremely low-level, low-frequency signal cables which would likely be susceptible to interference induced from currents flowing on the conduit, type-2 rubber-jacketed flexible conduit is preferred. The rubber jacket reduces the likelihood of current flow by insulating the conduit from incidental grounding contacts. An example of flexible conduit properly installed in a ship topside is shown in Figure 4-19.

Figure 4-19(a) Flexible Conduit Shielding Topside Cable Runs

Figure 4-19(b) Flexible Conduit Shielding Topside Cable Runs

To determine the correct conduit size, a general rule of thumb is that if a cable's outer diameter approaches 90 percent of a conduit's inner diameter, the next larger size conduit should be selected. For shipboard application, nominal sizes of conduit inner diameters range from ¼-inch up to 3 inches.

Rectangular metal troughs or trunks are frequently used in lieu of several individual conduits to provide EMI protection for large groups of nested cables in ship topsides. As shown before (a) and after (b) in Figure 4-20, when cables running up a mast cannot be placed inside a mast leg or center pole, they are enclosed within an EMI trunk.

Figure 4-20(a) Enclosing Mast Cable Runs Within EMI Trunk

Figure 4-20(b) Enclosing Mast Cable Runs Within EMI Trunk

In summary, the best methods to obtain sufficient shipboard cable shielding are choosing the correct cable from the requirements of MIL-C-24640 and -24643, eliminating common mode grounds, and carefully spacing between cable runs. EMC between the cables and suppression of EMI in the cables should be achieved by close adherence to established naval installation procedures such as those illustrated in Figure 4-21. Only when these practices cannot be followed adequately should the addition of shielding conduit be considered.

f. *RAM Shielding*—The problem of reflected energy was included in the earlier discussion of shipboard EMI sources. It was pointed out that, due to the complexity of a ship topside, the likelihood that radiated power will be unintentionally reflected from one or more metallic surfaces or deck objects is high. Depending upon the radiation frequency, reflections may be picked up as interference by such onboard receivers as navigation radars, search radars, missile tracking radars, weapons firing radars, TACAN, direction finder sets, and EW systems. The receptions appear as false targets or erroneous indications (i.e., bearings) which prompt inappropriate system reactions. Similarly, because of the congested topside environment and horizon-to-zenith 360° omnidirectional mission requirements, a high probability exists for direct path mainbeam or sidelobe RF coupling between shipboard emitters and sensors, resulting in severe interference and system degradation.

To a large extent the EMI potential for specularly reflected (and multiply scattered) energy and direct coupled RF can be mitigated by optimum placement of electromagnetic systems in the ship topside. (The engineering nature of topside design will be taken up later in this chapter.) Nevertheless, EMI due to reflected and coupled radiation does occur, oftentimes unpredicted and unexpected, during fleet operations. In answer to such occurrences, SEMCIP teams are dispatched to assist the ships in relieving the problems. In many cases the best solution is to employ RAM for EMI shielding.

Radar energy absorbers are specially devised materials which, due to their carefully contrived electromagnetic properties, have the ability to radically attenuate RF radiation. These materials have been designed for a wide variety of applications from 30 MHz through 100 GHz, but are most practical and effective at microwave frequencies. The concept is not new; experiments with RAM originated a half-century ago. Moreover, for the past 30 years RAM has been used routinely by the British navy to reduce false echoes from radar reflections off ship masts and superstructure by as much as 30 dB. Only in the last few years, however, has RAM begun to gain widespread application by the US military for suppression of EMI and for radar cross-section (platform image) reduction.

There are two principal types of electromagnetic energy absorbers used as RAM: narrowband resonant (tuned) attenuators and broader band graded dielectric attenuators. Resonant-type RAM is preferred in general for shipboard use because of its superior durability at sea and its comparative thinness. Resonant absorbers have developed over the years from the early Salisbury screen, a simple free space (377-ohm), thin, resistive sheet spaced precisely one–quarter wavelength from a conductive plane. Wave energy impinging upon the Salisbury screen is partially reflected at the screen surface and partially transmitted to the conductive rear plane. Upon meeting the conductive plane, the transmitted portion undergoes a series of multiple reflections. Part of each reflected vector is retransmitted outward, parallel to but 180° out of phase with the original surface reflected wave. As a result the vector sum of the multiple retransmissions is, in theory, equal to the original surface reflected energy, but, being 180° out of phase, effectively cancels it so that the total reflection is zero. Therefore, the Salisbury screen, at a very narrow frequency where the spacing is exactly a quarter wavelength, seemingly absorbs all the resonant incident energy; hence the term absorbing material. In practice, complete cancellation of the incident RF energy is never realized; nonetheless, attenuation of up to 30 dB (99.9% suppression) is achievable.

LIST OF MATERIAL

ITEM NO	PART	SPECIFICATION	NOTE
1	CONDUIT, SHIELDING	COMMERCIAL	3
2	CONDUIT ADAPTER, THOMAS & BETTS	COMMERCIAL	2,3
3	BACKSHELL 90°	MS-3188B-21A	3
4	BACKSHELL, 90°	MS-3188B-24A	3
5	ADAPTER, METAL BODY	COMMERCIAL	2
6	WEATHERSEAL, SIGMAFORM, SERIES SYT/H	COMMERCIAL	7

APPLICABILITY

EQUIPMENT	LCTN
LRA (P/S) ALL CABLES	FR 36 (P/S)
LRA (P/S) ALL CABLES	FR 36 (P/S)
STBD LRA, P7 ONLY	FR 36 (P)
LRA (P/S) P6 & P8	FR 36 (P/S)
LRA (P/S) ALL CABLES	FR 36 (P/S)
LRA (P/S) ALL CABLES	FR 36 (P/S)

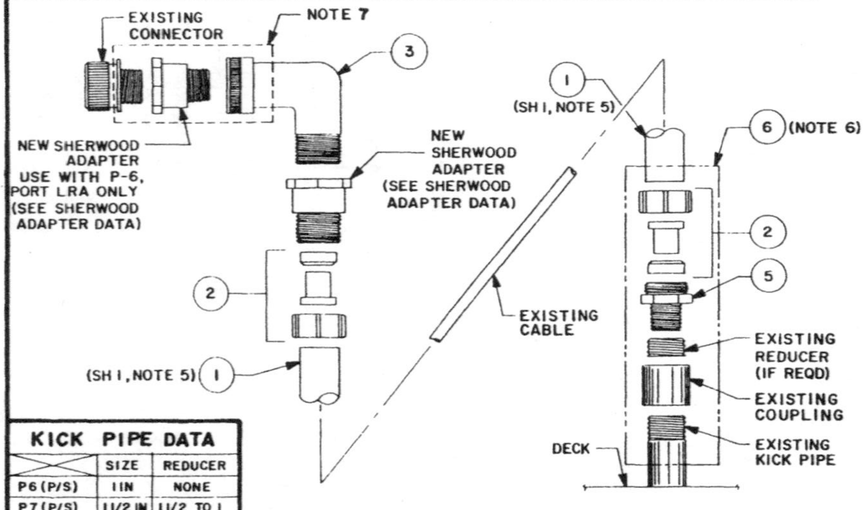

KICK PIPE DATA

	SIZE	REDUCER
P6 (P/S)	1 IN	NONE
P7 (P/S)	1 1/2 IN	1 1/2 TO 1
P8 (P/S)	1 1/2 IN	1 1/2 TO 1
P9 (P/S)	1 IN	1 TO 1/2
P10 (P/S)	1 IN	1 TO 1/2

HARPOON LAUNCHING RELAY ASSEMBLY
SHIELDING CONDUIT AND FITTINGS DETAIL

NOTES:

1. REMOVE CONNECTOR FROM TERMINATING EQUIPMENT AND DISCONNECT CABLE FROM CONNECTOR. REMOVE AND DISCARD PACKING GLAND NUT.
2. SELECT AN ADAPTER (ITEM 5) TO FIT COUPLING (OR REDUCER) AND THOMAS AND BETTS CONDUIT ADAPTER (ITEM 2). COAT LOWER ADAPTER THEADS WITH MIL-T-22361 ANTI-SEIZE COMPOUND OR KOPR SHIELD AND SCREW ADAPTER INTO COUPLING IN PLACE OF GLAND NUT AND TIGHTEN AS REQD FOR PACKING.
3. SELECT BACKSHELL, SHERWOOD ADAPTER AND UPPER CONDUIT ADAPTER TO FIT CONDUIT AND EXISTING CONNECTOR. CUT CONDUIT TO FIT AND PULL CABLE THRU CONDUIT AND FITTING AS SHOWN IN DETAIL.
4. COAT ALL THREADS WITH MIL-T-22361 ANTI-SEIZE COMPOUND OR KOPR SHIELD AND ASSEMBLE FITTINGS AND CONDUIT AS SHOWN.
5. RECONNECT CABLE TO CONNECTOR AND ENSURE THAT METAL TO METAL CONTACT IS MAINTAINED FROM THE CONNECTOR TO THE KICK PIPE. NOTE THAT PLUGS ARE ISOLATED, THEREFORE SHIELD IS GROUNDED AT DECK PENETRATION ONLY.
6. INSTALL ITEM NO. 6 AS SHOWN.
7. WEATHER SEAL IAW SH 1, NOTE 4 AND E18896.
8. WHEN ORDERING SHERWOOD ADAPTERS SPECIFY BACKSHELLS (MANUFACTURER & PART NO.) AND CONDUIT SIZE.

SHERWOOD ADAPTER DATA (NOTE 8)

PART NO.	SIZE	APPLICABILITY
MIL-C-5015/24-1.75-18-2-100-B	1 IN	P6 (P/S)
MIL-C-5015/21-1.437-18-2-100-B	1 IN	P7 (P/S)
MIL-C-5015/24-1.187-18-2-100-B	1 IN	P8 (P/S)
MIL-C-5015/16-.875-20-2-50-B	1/2 IN	P9 (P/S)
MIL-C-5015/16-.875-20-2-50-B	1/2 IN	P10 (P/S)
MIL-C-5015/SE504-2.06-16L-2-2-18-B	1 IN	P6 (PORT ONLY)

NAVSEA DRAWING NO.	REV	SHEET
806-6215591	A	13

Figure 4-21(a) Shipboard Cable Terminal Assembly Details

LIST OF MATERIAL

ITEM NO.	PART	SPECIFICATION	NOTE
1	GROUNDING ADAPTER (SIGMAFORM)	COMMERCIAL	4
2	SLEEVE, CABLE REPAIR (SIGMAFORM)	MIL-I-23053/15	10
3	GASKET MATERIAL, CONDUCTIVE	COMMERCIAL	2
4	SCREW, 1/8 x 20 UNC-2A x	CRES TYPE 316	
5	NUT, 1/8 x 20 UNC-2B	CRES TYPE 316	
6	SCREW, 1/4 x 20 UNC-2A x 3-3/4	CRES TYPE 316	
7	NUT, 1/4 x 20 UNC-2B	CRES TYPE 316	

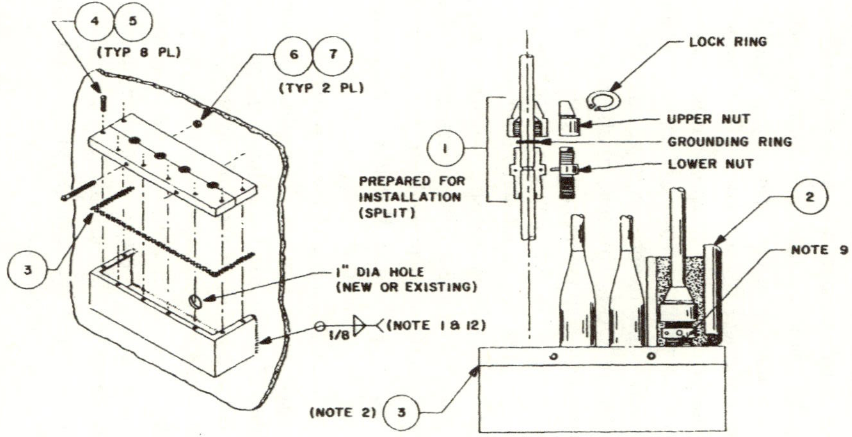

COAXIAL CABLE BHD PENETRATION BOX AND SHIELDING ADAPTER INSTALLATION DETAILS

NOTES:

1. INSTALL ONE (1) BULKHEAD PENETRATION BOX FOR COAXIAL CABLES AT AFT PILOT HOUSE BHD, APPROX 2 METERS STBD OF CL, AND 2 METERS ABOVE 01 LEVEL (WEATHER DECK). SEE SHEET 19 FOR FABRICATION DETAILS. SEE NOTE 12.
2. INSTALL EMI GASKET BETWEEN COVER AND BOX AS SHOWN AND IAW SH 1, NOTE 5.
3. ROUTE COAXIAL CABLES INDIVIDUALLY THROUGH APPROPRIATE HOLES IN COVER AND THROUGH NEW OR EXISTING HOLE IN BULKHEAD, AND ASSEMBLE BOX AND COVER AS SHOWN.
4. SELECT GROUNDING ADAPTER TO FIT HOLES IN COVER AND DISASSEMBLE FOR INSTALLATION.
5. REASSEMBLE LOWER NUT AROUND CABLE. COAT ALL THREADS WITH AN ANTI-SEIZE COMPOUND OF MIL-T-22361, OR KOPR SHIELD, THREAD NUT INTO HOLE IN COVER, TIGHTEN AS REQUIRED FOR PACKING.
6. USING A POCKET KNIFE OR SIMILAR TOOL, CUT THE CABLE JACKET AROUND THE PERIMETER OF THE CABLE FLUSH WITH THE TOP OF THE LOWER NUT. MAKE ANOTHER CUT IN THE CABLE JACKET APPROXIMATELY ONE-FOURTH INCH HIGHER AND REMOVE THE CUT SECTION OF CABLE JACKET.
7. COAT THE EXPOSED CABLE BRAID WITH THE ANTI-SEIZE COMPOUND IDENTIFIED IN NOTE 5.
8. ASSEMBLE GROUND RING AND UPPER NUT AROUND CABLE. LOCK UPPER NUT SECTIONS TOGETHER WITH SNAP RING. PLACE THE GROUNDING RING AROUND THE CABLE WHERE JACKET WAS REMOVED AND HAND-TIGHTEN UPPER NUT TO LOWER NUT.
9. COAT LOWER ADAPTER THREADS, SEAM IN COVER, BULKHEAD EDGE OF COVER AND GASKET EDGES WITH "PERMATEX II", "SCOTCHCLAD", OR "BMS-595".
10. PLACE CABLE REPAIR SLEEVE AROUND COMPLETE ADAPTER AND HEAT SHRINK AS REQUIRED FOR WEATHER SEALING, ENSURING AREA AROUND CABLE IS THOROUGHLY SEALED.
11. PLUG ANY UNUSED HOLES AND ENSURE THAT BOX AND COVER ARE COMPLETELY WEATHER SEALED.
12. SURFACES AFFECTED BY THIS INSTALLATION SHALL BE CLEANED AND PAINTED IAW NAVSEA S9AA0-AA-SSPN-010, SECTION 631 OF GENERAL SPECIFICATIONS FOR SHIPS OF THE U.S. NAVY.

NAVSEA DRAWING NO.	REV	SHEET
806-6215591	A	18

Figure 4-21(b) Shipboard Cable Terminal Assembly Details

Contemporary technical improvements to the original Salisbury screen have produced resonant absorbers which are thinner, more flexible, more adaptable, and better resistant to weather. Furthermore, by impregnating them with high-permeability powdered ferrites, modern resonant absorbers offer high attenuation characteristics in addition to the destructive interference from multiply reflected out-of-phase components.

The three major types of tuned frequency resonant absorbers today are: magnetically loaded solid RAM; dielectrically loaded solid RAM; and sandwich-layered combination RAM. At discrete frequencies narrowband RAM can be designed for optimum performance as a function of material thickness, material composition, ingredient attenuation factor, and surface impedance. Also, resonant-type RAM can be made to perform well over multiple (normally two-band) frequencies by proper selection of critical magnetic and dielectric loading along with layer thickness.

Most tuned absorbers use lossy elastomeric material bonded to a conductive metal backing. Neoprene is often recommended for shipboard applications because of its high resistance to the sea environment; however, a variety of other base elastomers are available, including natural rubber, silicone, polyisoprene, nitrite, and urethane. When installed aboard ship such RAM is usually in the form of thin flexible sheets applied directly to the metal surfaces by adhesives. To improve its weather resistance, RAM should be covered with epoxy or urethane-based paints.

In contrast to the narrowband resonant absorbers using the quarter-wave destructive interference principle, there is an alternative broadband graded dielectric RAM. This type of absorber incorporates a varying material impedance which tapers gradually from a free space surface through a lossy interior to a low impedance rear surface. The impedance transition must be achieved slowly to allow as little reflection as possible to escape. This is done by combining precise geometrical shaping (e.g., the pyramidal configuration used for anechoic chamber surfaces) with a conductive carbon filler in a polyurethane foam medium. Using such techniques, attenuation as high as 50 dB is attained. Because of the requirement for very gradual transition in impedance from front surface to back, however, this type of RAM tends to be thicker than the resonant absorbers, and, because they are foam-based, they are relatively fragile. Consequently, graded dielectric RAM is not frequently used in shipboard applications.

All RAM types are susceptible to physical and electromagnetic degradation aboard ship because of continuous exposure to sunlight, water, salt, exhaust contaminants, oil, ice, heat, corrosion, high winds, and such maintenance practices as use of metal-based paints and overpainting. Yet, if properly cared for, the useful life expectancy of shipboard RAM is currently about eight years. Some of the environmental factors that adversely affect the performance of RAM are:

1. *Water*—A film of water of any kind (rain water and sea spray are utterly commonplace in shipboard topsides) severely reduces the effectiveness of RAM. Special coatings which facilitate water run-off or beading help to restore good performance. Salt water film must be washed off with fresh water. Once the RAM surface has dried, full effectiveness is restored.
2. *Paint*—Metal-based paints have a deleterious effect upon RAM performance, especially when several coats have accumulated. Standard Navy paints use titanium dioxide as a primary pigmentation ingredient. Experience in the fleet indicates that up to three coats of standard haze gray paint can be applied before seriously affecting RAM quality. Performance is destroyed with six or more coats. The results are worse with lead-based paints, and, of course, iron-based types cause rusting. Polyurethane nonmetallic paints are recommended for protective coatings of shipboard RAM.
3. *Salt*—Coatings of salt cause tuned absorber-type RAM to shift downward in resonant frequency. The thicker the salt coating, the more the shift, so that the effective performance of the RAM is significantly degraded. Washing the salt off with fresh water and drying restores full original quality.
4. *Ice*—Coatings of ice also deteriorate RAM performance, but not as severely as salt coatings. When the ice is melted off and the surface dried, full performance is restored.
5. *Rust*—RAM that uses ferrite composites for attenuation must be painted when installed aboard ship to prevent rusting. If a thin film of rust develops, the surface must be cleaned thoroughly with a mild acid, dried, and painted. If rusting is allowed to continue, the elastomer base will crumble and delaminate so that the RAM must be replaced.
6. *Oil*—Rubber elastomers used for most resonant-type RAM are vulnerable to petroleum products. Oil, gasoline, and related chemicals cause an elastomer to soften and deteriorate rapidly, so it must be washed off immediately with mild detergent and water. If the rubber shows evidence of disintegration, the RAM must be replaced. Most shipboard tuned RAM has neoprene rubber as the base, which is most resistant to petroleum degradation.
7. *Sunlight and Air*—Over a long period of time, generally from four to six years, exposure to sunlight and atmosphere will cause RAM elastomers to crack and crumble. If deterioration is allowed to progress, flaking occurs and the RAM must be replaced. Painting of the RAM retards deterioration caused by long-term exposure to sunlight and air.
8. *Heat*—Even though most RAM will withstand temperatures from −65°F to 250°F, excessive heat will blister, melt, and even burn RAM. After any such occurrence, the RAM must be replaced.

Bonded to masts, stacks, yardarms, bulkheads and other reflective surfaces, or erected as rigid RF barriers, RAM has proven to be highly useful for attenuating reflected energy and blocking direct coupled EMI. In recent applications aboard naval ships, RAM has been:

- Installed on aircraft carrier yardarm structures to reduce reflected EMI causing false images in air control radar displays.
- Applied to the inside of weapons director tubs to prevent false target lock-on of the directors from reflected energy.
- Erected as EMI barriers to block reflected energy from being received as false emissions by ECM systems on aircraft carriers (see Figure 4-22).
- Applied to battleship air search radar pedestals to reduce a large blind spot seen by nearby lower surface search radar (see Figure 4-23).

Figure 4-22(a) RAM Barriers Protecting Aircraft Carrier ECM System

Figure 4-22(b) RAM Barriers Protecting Aircraft Carrier ECM System

Figure 4-23 RAM Coating on Air-Search Radar Pedestal to Reduce Blind Spot

- Wrapped around tall HF monopole whip antennas near weapons control radar directors to eliminate reflected interference on cruisers.
- Wrapped around large mast legs on amphibious assault ships to reduce structural reflections picked up by surface search and air search radars (see Figure 4-24).
- Used to coat flag bags on amphibious assault ships to alleviate reflected interference received by electronic warfare sensors.
- Erected as EMI barriers on the Fresnel lens assemblies of aircraft carriers to prevent reflected energy from causing severe degradation of electronic warfare systems (see Figure 4-25).

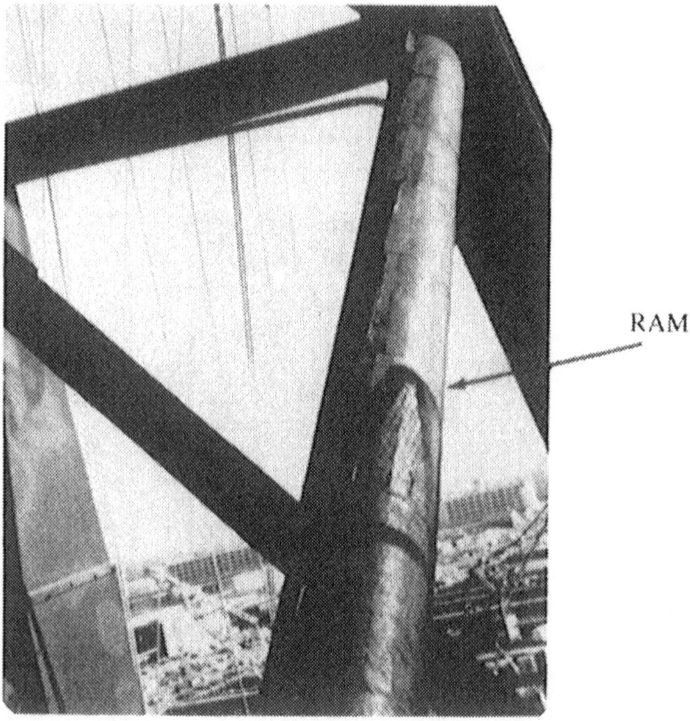

Figure 4-24 RAM Wrapping on Mast Legs to Reduce Reflective Surfaces

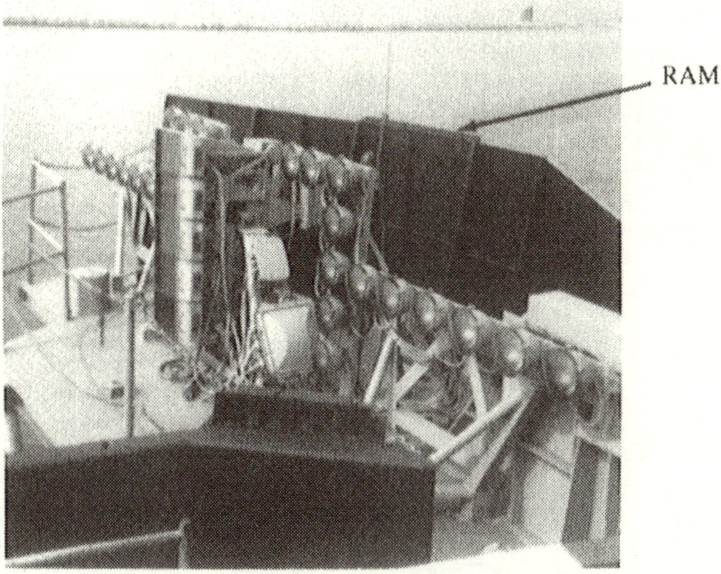

(a)

(b)

Figure 4-25 RAM Barriers to Shield Fresnel Lens EMI on Aircraft Carriers

- Applied to aircraft carrier auxiliary conning stations to prevent RF energy reflections from interfering with electronic countermeasures sensors (see Figure 4-26).
- Installed on the UHF satellite antenna support structures of destroyers to prevent the reflecting and scattering of RF energy (see Figure 4-27).
- Erected as an EMI barrier to prevent direct coupling of radiated energy between electronic warfare systems and SHF satellite communications antenna on amphibious command ships (see Figure 4-28).

Figure 4-26 RAM Coating of Aircraft Carrier Auxiliary Conning Station to Prevent RF Reflections

Figure 4-27 RAM Installed on Antenna Support Platform to Reduce Reflections

Figure 4-28 RAM Barrier to Block EMI Between EW and SHF SATCOM Systems

In summary, the use of RAM is becoming increasingly relied upon as a solution to EMI problems in the topsides of naval warships. However, at present there is no Navy standard RAM. Instead, commercial types of RAM are sought or naval laboratory experimental materials are fabricated to be applied as custom-made fixes for individual cases. Finally, bear in mind that the broader the frequency range of application and the lower the frequency, the thicker, the heavier, and the more expensive the RAM.

4-2.2 Grounding and Bonding Techniques

4-2.2.1 General Definitions

The Navy has long recognized that proper grounding and bonding is essential to control shipboard EMI effectively. Hull-generated "rusty bolt" intermodulation products in particular would proliferate without careful bonding of nonlinear junctions. Likewise, faulty grounding of cable shields and connectors would allow cables penetrating the ship interior from topside areas to transport EMI surreptitiously straight into susceptible equipment. Correct application of grounding and bonding is considered of paramount importance to naval electromagnetic systems engineering, installation, and maintenance.

But just what is meant by grounding, and how does grounding differ from bonding? There is a definite distinction between the two, even though the difference may, until well understood, seem superficial. Grounding is a necessity aboard ship foremost to protect personnel from electrical shock hazards. Secondarily, it is an important means of suppressing EMI, especially at low (e.g., powerline and audio) frequencies. In essence, grounding is an electrical circuitry practice; i.e., it is integral to the ship's overall electric system. Its purpose is to establish and maintain near zero-resistance conductive paths to a common reference point ground. By connecting all electrical devices to this common reference point, or ground plane, there is (ideally) zero potential difference between all connected points anywhere in the electrical system. Having no potential difference eliminates the possibility of electrical hazards to personnel coming into external contact with the various system components. It also has the added benefit of conducting many types of EMI directly to ground. The reference ground plane of a metallic ship is the hull itself and all structures bonded to the hull, by virtue of the hull contact with sea water.

Of course, an ideal zero-potential ground network cannot be realized in actual practice, where, as on a ship, the total electronic and electrical system is so complex. Reality notwithstanding, zero-potential is closely approached at dc and very low frequencies, and gradually becomes less ideal at higher frequencies due to the innate capacitive and inductive characteristics of the ground system wiring and components. To minimize these reactive characteristics, grounding lead lengths are kept as short as possible by multipoint grounding; that is, by grounding directly to the closest available point of the ground plane. Even so, multipoint grounding may introduce so-called ground loops, from which differences in potential cause EMI currents to flow, diluting the ground system quality. Therefore, care must be taken to avoid the setting up of ground loops during the installation of shipboard equipment and in the carrying out of grounding practices.

Bonding, like grounding, establishes a highly conductive electrical path, and in that sense is, in fact, a special form of grounding. As contrasted to grounding, however, bonding: (1) is principally an electromagnetic RF mechanism, and (2) provides a low-impedance path across a single junction, mated surfaces, or pairs of metallic elements, without regard to the overall ship electrical system. As such, bonding has as its chief purpose the elimination of potential sources of intermodulation noise. Because we are more interested in electromagnetic principles and the control of shipboard EMI, it is bonding practices to which we will devote most of our emphasis.

In a perfect zero-impedance shipboard electromagnetic environment, in which all components and extensions of the hull create a single electrically continuous equipotential ground system, there would be no sources of intermodulation noise products. In such a perfect system all metallic junctions between the hull and exterior members, structures, appendages, and surfaces would be at equipotential.[24] We have seen in Section 4-1.2.d., however, that in the real world there are very many potential nonlinear junction intermodulation creators aboard naval ships. Detecting and suppressing these RF noise generators is an unending engineering effort.

4-2.2.2 Bonding Classifications

Four techniques are used to quell intermodulation interference:

a. Replace the offending shipboard source element with one made of nonmetallic material.
b. Completely insulate various potential source components from each other and from the hull.
c. Wherever possible, bond all hull structure junctions by welding or brazing.
d. Attach bond straps across junctions that cannot be welded or brazed.

The first two of these techniques will be taken up later in this chapter. It is bonding methods that are of interest now.

The authoritative document for all grounding and bonding aboard US naval ships is MIL-STD-1310, which describes items to be bonded, bond strap fabrication, installation details, and attachment methods. Highlights of this document will be presented here.

Shipboard bonding methods are classified as follows:[25]

a. *Class A*—A bond established by joining two metallic surfaces through welding or brazing.
b. *Class B*—A bond achieved by metal-to-metal contact through normal installation of equipment using mounting bolt hardware.
c. *Class C*—A bond established by bridging two metallic surfaces with a metallic strap.

Bonding by welding or brazing (i.e., class A bonding) is called direct bonding, and in all events is the preferred method. Properly done, class A bonding results in the lowest impedance union and requires the least maintenance. However, there are many instances where a piece of shipboard equipment or system component simply cannot be fixed permanently in one position. Hatch covers, for example, must open and close; mechanical linkages must be allowed freedom of movement; and many types of electronic equipment must be installed on shock mounts. In such cases indirect bonding using bolts (class B) or jumper straps (class C) must be resorted to. In no case should riveting or self-tapping screws be used to establish a bond. When employing a bond strap, the strap must be kept as short as at all possible. Flat types must never exceed a length-to-width ratio of 5. Also, bear in mind that, although bond straps do initially form a low-impedance union, they gradually become less effective because of corrosion. Exposure to the weather promotes hasty deterioration unless sufficient surface preservation is maintained. If the deterioration is allowed to continue, the corroded bond attachment itself may become a generator of EMI.

Metallic straps used as class C bonds are categorized as follows:

a. *Type I*—A strap made of TRXF-84 or equal flexible welding cable with terminal lugs of steel or aluminum, used in shipboard topside areas to bond across intermodulation interference sources or to bond equipment to the hull, where class A bonding is inapplicable. Type I straps are waterproof, very flexible, and well-suited for the harsh shipboard environment. The type of lug selected for the strap must match the mating surface; i.e., steel lugs for attachment to steel surfaces, and aluminum lugs for aluminum surfaces. One end is welded to the ship hull ground and the other end welded to the item requiring bonding. The bond strap length must be selected on the basis of the particular bonding requirement, always using the shortest length possible. Type I bond straps are normally fabricated in 6-, 9-, or 12-inch lengths. See Figure 4-29 for details.

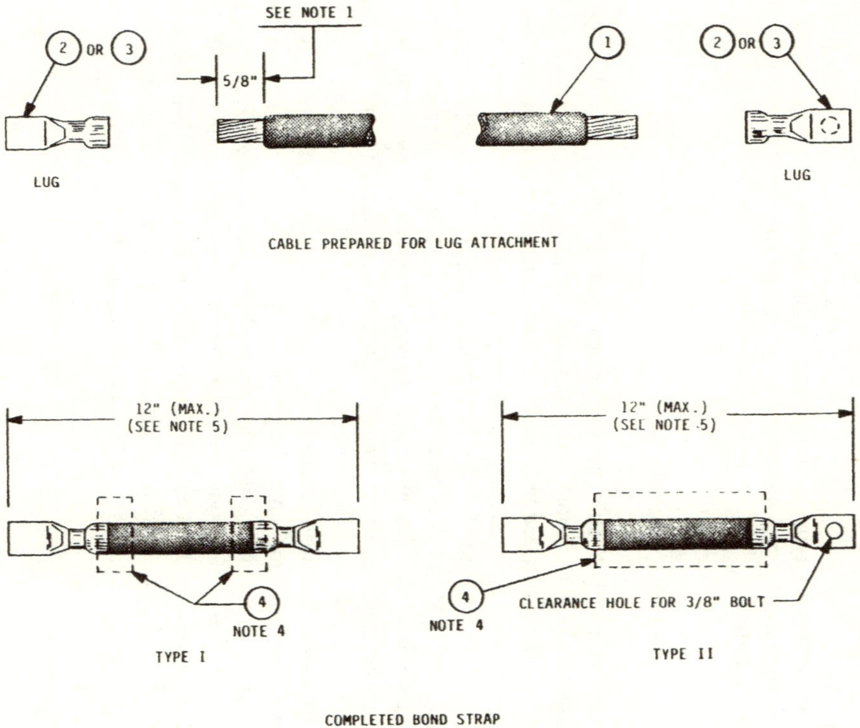

LIST OF MATERIAL

ITEM NO.	PART	SPECIFICATION	NOTE
1	CABLE, WELDING, TRXF-84	MIL-C-915/21	1,2,3,4
2	LUG, ALUMINUM	COMMERCIAL	1,2,3
3	LUG, CRES 316	COMMERCIAL	1,2,3
4	TUBING, HEAT-SHRINKABLE	MIL-I-23053/15	4

NOTES:

1. CABLE JACKET SHALL BE REMOVED FIVE-EIGHTS INCH OR AS APPROPRIATE TO ALLOW THE EXPOSED COPPER CONDUCTORS AND THE JACKETED PORTION OF THE CABLE TO PENETRATE TO THE FULL DEPTH OF THEIR RESPECTIVE CHAMBERS IN THE LUG. ENSURE THE CONDUCTORS ARE BRIGHT COPPER.

2. LUG AND BOND STRAP ASSEMBLY TOOLS SHALL BE THE FOLLOWING PART NUMBERS, OR EQUAL:

   THOMAS & BETTS CO., RARITAN, NJ:

LUG, CRES 316	(WITH PUNCHED TONGUE)
LUG, CRES 316	(BLANK TONGUE)
LUG, ALUMINUM	271-30483-334
12-TON HEAD	13642
DIE, FOR STEEL LUG	297-54251
DIE, FOR ALUMINUM LUG	297-53346
HYDRAULIC HOSE	21089
HYDRAULIC PUMP	13606

   RESEARCH TOOL & DIE WORKS, CARSON, CA:

LUG, CRES 316	SST-8315 (PUNCHED TONGUE)
LUG, CRES 316	SST-8316 (BLANK TONGUE)
LUG, ALUMINUM	SSA-8315 (PUNCHED TONGUE)
LUG, ALUMINUM	SSA-8316 (BLANK TONGUE)

3. LUG MATERIAL, STEEL OR ALUMINUM, SHALL BE SELECTED TO MATCH THE CORRESPONDING MATING SURFACE MATERIAL. THE DIE ASSEMBLY SHALL BE COMPRESSED TO CLOSURE TO PROPERLY CRIMP THE LUGS ONTO THE CABLE.

4. EACH BOND STRAP SHALL BE INSPECTED AFTER COMPLETION BY SLIGHTLY BENDING THE BOND STRAP AT EACH TERMINAL LUG AND LOOKING FOR A GAP BETWEEN THE CABLE JACKET AND THE LUG SHROUD. WHERE WEATHERSEALING IS NOT APPARENT, HEAT-SHRINKABLE TUBING SHALL BE ADDED AS SHOWN. TUBING SHALL BE SHRUNK ONLY AFTER BOND STRAP HAS BEEN WELDED IN PLACE SINCE HEAT FROM WELDING WOULD LOOSEN THE TUBING WEATHERSEAL.

5. IN CERTAIN CASES, SUCH AS ON LIFE NETS, BOND STRAPS MAY BE REQUIRED TO BE SLIGHTLY LONGER TO PREVENT BREAKING THE BOND STRAP WHEN THE NETS ARE RAISED OR LOWERED. BOND STRAPS INSTALLED ON CLIMBER SAFETY RAILS MAY ALSO REQUIRE A SLIGHT INCREASE IN LENGTH.

Figure 4-29 Type I and Type II Bond Strap Details

b. *Type II*—A strap identical to type I except that one lug has a drilled or punched hole to accommodate a threaded stud or bolt. Type II straps are used to bond equipment or devices that cannot be permanently fixed in place, so that only one end of the strap is welded, and the other is bolted down. Use of type II bonding must be kept to a minimum. See Figure 4-29 for details.
c. *Type III*—A flat, solid copper strap for use in topside areas or below decks for bonding such items as antenna tuners and couplers, equipment enclosures, and cabinets. These straps are normally available in 3-, 6-, 9-, and 12-inch lengths. See Figure 4-30 for details.
d. *Type IV*—A flat, braided copper strap for bonding sound-isolated mounts and for bonding electromagnetic shielding conduit aboard submarines. Normally available in 3-, 6-, 9-, and 12-inch lengths. See Figure 4-30 for details.

Of the four types, the flat straps offer the least RF impedance, the braided straps have the highest flexibility, and the wire cable straps are the least expensive. However, the type of bond strap to be employed is selected more in accordance with the particular situation or circumstance than by such factors as cost or flexibility.

4-2.2.3 Grounding Requirements

On a metal-hull ship the designated ground reference is the hull. All equipment racks, foundations, structures, and other large metal items are welded, brazed, or class-C-bonded by a low-resistance connection to the hull to become, by extension, the same ground reference potential.

The basic criterion for electrical protection and EMI reduction aboard ship is that all electrical and electronic equipment and workbenches must be grounded. Equipment installed on resilient mounts must be grounded by a third conductor in the power supply cable or bonded to ground as shown in Figure 4-31. Equipment not installed on resilient mounts is considered properly bonded by metal-to-metal contact and installation bolts (i.e., class-B-bonded). Slide-mounted or roller-mounted equipment must be grounded by a conductor within the equipment cable harness. If a ground conductor has not been provided by the manufacturers or installers, a flexible ground conductor must be installed between the drawer frame or chassis and the enclosure frame at ground potential. The ground conductor size must be equal to or greater than the size of one of the ac–power conductors supplying power to the drawer equipment.

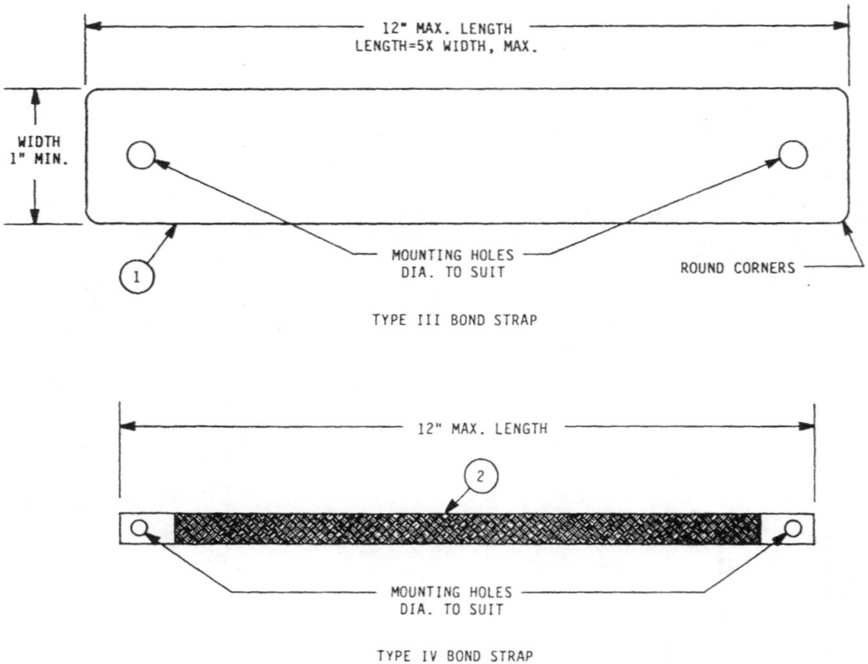

LIST OF MATERIAL			
ITEM NO.	PART	SPECIFICATION	NOTE
1	FLAT COPPER	QQ-C-576	1
2	FLAT COPPER BRAID	QQ-B-575	2

NOTES

1. TYPE III BOND STRAPS SHALL BE FABRICATED FROM FLAT COPPER NOT LESS THAN 0.020" THICK OF QQ-C-576, SHALL NOT EXCEED A LENGTH-TO-WIDTH RATIO OF FIVE-TO-ONE, AND SHALL BE ZINC PLATED PER ASTM-B-633. MOUNTING HOLES SHALL BE PROVIDED IN EACH END OF TYPE III BOND STRAPS.

2. BOND STRAP MATERIAL SHALL BE FLAT COPPER BRAID, 1.0 INCH MINIMUM WIDTH, IN ACCORDANCE WITH QQ-B-575. END TERMINALS SHALL BE CUT FROM FLAT COPPER OF QQ-C-576 AND SHALL BE 1.0 INCH WIDE BY 2.0 INCHES LONG. END TERMINALS SHALL BE HOT TIN DIPPED, THEN SOLDER COATED ON ONE SIDE. EACH TERMINAL SHALL BE BENT 180 DEGREES TO FIT OVER BRAID USING 1/16 INCH METAL PLATE AS A BENDING TEMPLATE. ENSURE SOLDER COATING IS ON INSIDE. BRAID MATERIAL SHALL BE FLUX COATED 1.0 INCH ON EACH END. END TERMINALS SHALL BE HEATED AND COMPRESSED ONTO THE BRAID USING ADDITIONAL SOLDER AS REQUIRED. THE REQUIRED HOLES SHALL BE PUNCHED IN EACH END.

3. WHERE A TYPE IV BOND STRAP IS USED ONLY FOR PERSONNEL SAFETY GROUNDING, THE LENGTH OF THE STRAP SHALL BE AS REQUIRED.

Figure 4-30 Type III and Type IV Bond Strap Details

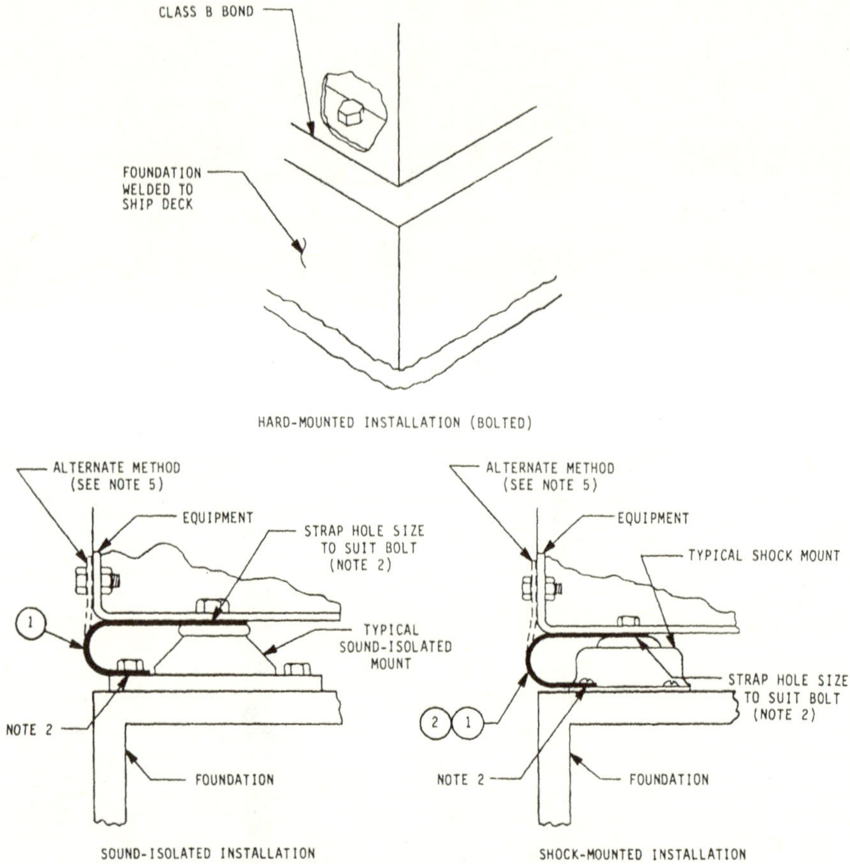

LIST OF MATERIAL			
ITEM NO.	PART	SPECIFICATION	NOTE
1	BOND STRAP, TYPE IV		1
2	BOND STRAP, TYPE III		3

NOTES:

1. BOND STRAPS FURNISHED WITH ELECTRONIC EQUIPMENT BY MANUFACTURERS MAY BE USED FOR BONDING RESILIENT MOUNTED EQUIPMENTS IF EQUAL TO, OR SIMILAR TO, THE TYPE IV BOND STRAP AS SPECIFIED HEREIN. WHEN A BOND STRAP IS NOT FURNISHED WITH AN EQUIPMENT, A TYPE IV BOND STRAP SHALL BE INSTALLED. BOND STRAP INSTALLATION SHALL NOT DEFEAT PURPOSE OF RESILIENT MOUNT.
2. WHERE POSSIBLE, USE EXISTING BOLTS, STUDS, OR HOLES FOR ATTACHING BOND STRAP.
3. AS AN ALTERNATIVE, A TYPE III BOND STRAP MAY BE INSTALLED ON SHOCK-MOUNTED EQUIPMENT.
4. EACH BOND STRAP INSTALLED SHALL ACCOMODATE THE FULL DEFLECTION OF EACH RESILIENT MOUNT.
5. IF SHIMMING OF REMAINING MOUNTS IS REQUIRED DUE TO THE INSTALLATION OF THE BOND STRAP, THE ALTERNATIVE GROUNDING METHOD MAY BE USED.

Figure 4-31 Grounding of Equipment Enclosures

Metal-cased portable electrical equipment and electronic test equipment must come equipped with, or be modified to use, three-wire, three-prong cable assemblies. Additionally, power cords having metal-covered plug assemblies must be replaced with molded plug power cable assemblies. Temporary and portable shelters such as huts, vans, and trailers that contain electrical and electronic equipment, that are located in weather deck areas, and that are not bolted to the ship deck (class-B-bonded) must be grounded by type II bond straps. Shelters equipped with antennas and requiring RF grounding must be grounded by type III bond straps. Tiedown cables should be nonmetallic, but if metallic cables are used, insulators must break the cable every five feet. Antenna tuners, couplers, matching networks, and receive termination boxes must be grounded by type III bond straps.

In each of the preceding grounding requirements, the dc resistance between each item or unit of equipment and the ground point must not exceed 1.0 ohm.

Because shipboard signal-carrying cables and RF transmission lines are likely to be either a source of EMI (radiator) or potentially susceptible to RF fields (susceptor), special precautionary grounding methods must be applied. All cables routed in topside areas must be shielded when possible. Unshielded cables should be placed within the ship structure or other metal enclosures such as masts. Coaxial cables and other types having a metallic sheath are considered properly shielded; the overall shield must be correctly terminated (360° grounding) at each terminal piece of equipment however, and must be grounded at weather deck penetrations. All other cables must be routed within rigid conduit, flexible conduit, or covered wireways. Examples of flexible conduit shielding are detailed in Figure 4-32 and must terminate in 360° grounding configurations as shown. Covered wireways, illustrated in Figure 4-33, may contain shielded and unshielded cables. Shielded cable in the wireway trunk must have an overall cable shield grounded as shown, using the methods of Figure 4-34. Unshielded cable in the wireway trunk must employ add-on shielding such as rigid or flexible conduit to the weather-exposed cable, properly grounded as depicted in Figure 4-32. Waveguides, pipes, and metal tubing routed in topside areas and penetrating a weather deck or bulkhead must be grounded at this point using the methods of Figure 4-35 for waveguide grounding, and those of Figure 4-34 (typical cable shield grounding) for deck pipes and stuffing tubes. Pipes welded at penetration points are considered properly grounded.

There are some naval ships, e.g., minesweepers and patrol craft, that have nonmetallic hulls. In such cases, special ground systems must be devised. Ground plates are affixed to each side of the keel as closely as possible to the propellor structure to provide an earth ground connection in contact with the sea. The ground plates are made of ⅛-inch-thick 2- by 4-foot copper plates. A through-bolt is brazed to each plate to allow a connection terminal for the ground cable system, and a 1/0 AWG cable is fastened between the two ground plates. As direct as possible a 1/0 AWG cable is run from the ground plates to the radio

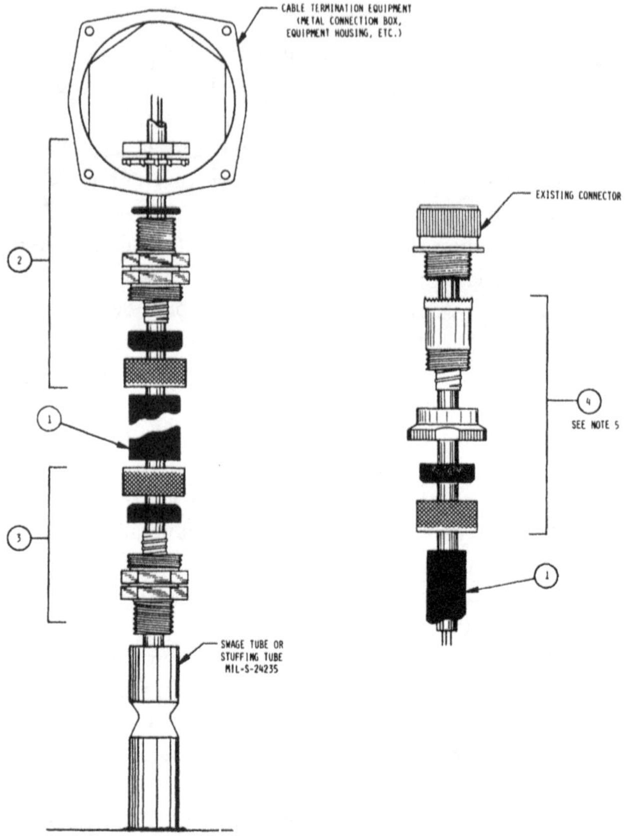

METHOD 1 - SHIELDING CONDUIT INSTALLATION, TYPICAL

LIST OF MATERIAL			
ITEM NO.	PART	SPECIFICATION	NOTE
1	CONDUIT, SHIELDING	COMMERCIAL	4,6
2	FITTING, CONDUIT-TO-BOX	COMMERCIAL	4,6
3	FITTING,		
	CONDUIT-TO-STUFFING TUBE	COMMERCIAL	3,4,6
4	FITTING, CONDUIT-TO-CONNECTOR	COMMERCIAL	4,5,6

NOTES:

1. THIS METHOD OF CABLE SHIELDING APPLIES TO NEW CABLE INSTALLATIONS AND TO EXISTING CABLE INSTALLATIONS WHERE THE CABLE CAN BE REMOVED FROM THE TERMINATING EQUIPMENT AND HANGERS AND ROUTED THROUGH THE CONDUIT.

2. FOR NEW CABLE INSTALLATIONS - PRIOR TO PULLING CABLE THROUGH STUFFING TUBE, REMOVE PACKING GLAND NUT FROM THE TUBE AND DISCARD.

   FOR RETROFIT INSTALLATIONS - REMOVE CABLE FROM TERMINATING EQUIPMENT (LIGHT, CONNECTION BOX, SPEAKER, CONNECTOR, ETC.) AND REMOVE CABLE FROM HANGERS DOWN TO STUFFING TUBE. UNSCREW PACKING GLAND NUT FROM STUFFING TUBE, REMOVE NUT FROM CABLE AND DISCARD.

3. FOR BOTH NEW AND RETROFIT INSTALLATIONS - SELECT CONDUIT-TO-STUFFING TUBE END FITTING TO MATCH CONDUIT SIZE AND TUBE SIZE. ROUTE CABLE THROUGH FITTING. PACK STUFFING TUBE (NEW INSTALLATIONS). COAT LOWER THREADS OF FITTING WITH ANTI-SEIZE COMPOUND OF MIL-T-22361. SCREW FITTING INTO STUFFING TUBE AND TIGHTEN AS REQUIRED FOR PACKING.

4. MEASURE AND CUT PROPER LENGTH OF CONDUIT TO COVER THE ENTIRE LENGTH OF CABLE FROM STUFFING TUBE TO END TERMINATION. ENSURE CONDUIT IS CUT SQUARE. FEED CABLE THROUGH CONDUIT. COAT UPPER THREADS OF FITTING WITH ANTI-SEIZE COMPOUND AND TERMINATE CONDUIT INTO FITTING AT STUFFING TUBE. INSTALL CONDUIT IN CABLE HANGERS AND TERMINATE OTHER END OF CONDUIT INTO END FITTING AT TERMINATING EQUIPMENT. ENSURE ALL METAL CONNECTING PARTS ARE COATED WITH ANTI-SEIZE COMPOUND PRIOR TO ASSEMBLY. CONNECT OR RECONNECT CABLE CONDUCTORS TO PROPER TERMINALS.

5. WHERE BOTH ENDS OF THE CONDUIT TERMINATE IN AN ELECTRICAL OR ELECTRONIC CONNECTOR, THE METHOD OF CONDUIT ATTACHMENT TO CONNECTOR MAY BE IN ACCORDANCE WITH MIL-C-28840.

(NOTES CONTINUE ON FOLLOWING PAGE)

6. SHIELDING CONDUIT AND ASSOCIATED END FITTINGS SHALL BE THE FOLLOWING, OR EQUAL. "OR EQUAL" SHALL BE DEFINED AS EQUAL IN ELECTROMAGNETIC SHIELDING, GROUNDING, CORROSION PROTECTION, AND WEATHERPROOFING.

   BREEZE-ILLINOIS, INC., WYOMING, IL,"BI-PRO 175" (TYPE SHOWN ON THIS FIGURE)

   ANAMET, INC., WATERBURY, CT, "SHIELDTITE"

   ETCON CORP., PALOS HEIGHTS, IL, "TYPE CC"

   GLENAIR, INC., GLENDALE, CA, "SERIES 75"

7. WEATHERSEAL CONNECTORS AND FITTINGS AS SPECIFIED FOR CORROSION PROTECTION.

Figure 4-32 Cable Shielding Methods

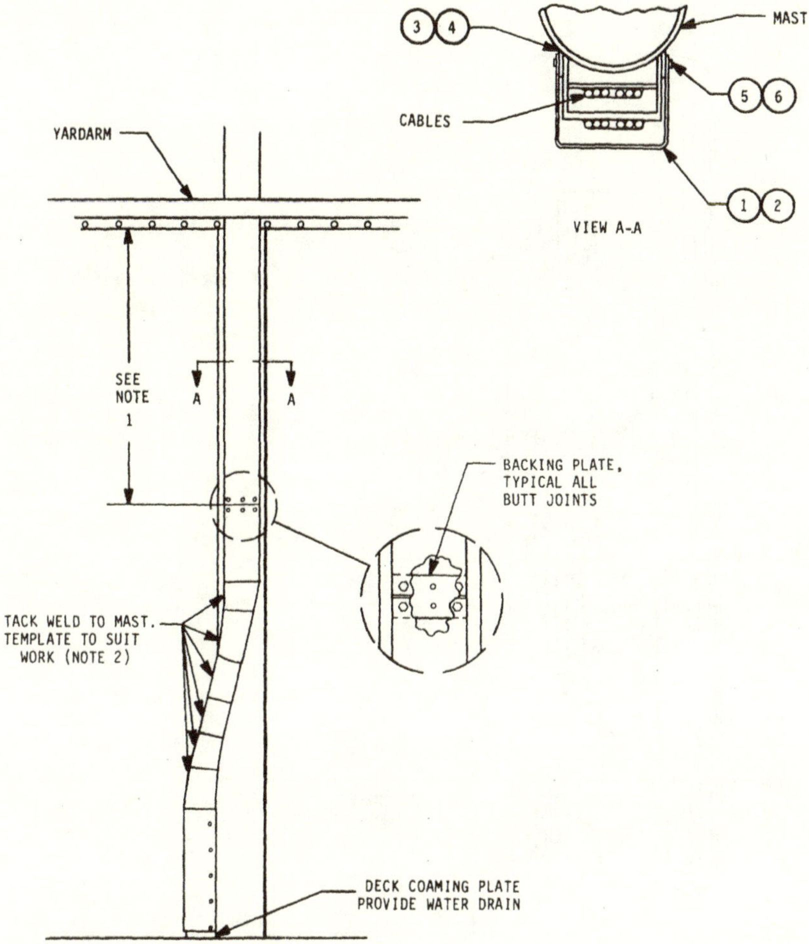

LIST OF MATERIAL			
ITEM NO.	PART	SPECIFICATION	NOTE
1	SHEET METAL, 1/16" GALV STEEL		2,3
2	SHEET METAL, 1/8" ALUM PLATE		2,3
3	FLAT BAR, 2"x1/4", STEEL		1
4	FLAT BAR, 2"x1/4", ALUM		1
5	BOLT, CRES	MIL-S-1222	
6	LOCKWASHER, CRES		

NOTES:

1. FLAT BAR AND WIREWAY TRUNKS SHALL BE ALUMINUM FOR ALUMINUM MASTS AND GALVANIZED STEEL FOR STEEL MASTS. DRILL AND TAP FLAT BAR AS REQUIRED. BAR SHALL BE TACK WELDED TO MAST FOR PROPER SUPPORT OF TRUNK. STRAIGHT TRUNK SECTIONS SHALL BE FABRICATED IN LENGTHS OF 8 FEET OR AS APPROPRIATE.

2. CURVED PORTIONS OF THE TRUNK RUN MAY BE FABRICATED IN SHORT STRAIGHT SECTIONS AS REQUIRED TO FACILITATE INSTALLATION OVER CURVED SECTIONS OF CABLE RUNS. THESE SHORT SECTIONS CAN BE TACK WELDED DIRECTLY TO THE MAST.

3. SECTIONS OF TRUNK (INSIDE AND OUTSIDE) SHALL BE PRIMED AND PAINTED. THE CONTACT SURFACE BETWEEN THE TRUNK AND FLAT BAR ON STEEL TRUNKS SHALL ALSO BE PAINTED. THE CONTACT SURFACE ON ALUMINUM TRUNKS SHALL NOT BE PAINTED AND SHALL BE TREATED WITH A CLASS 3 CONDUCTIVE COATING OF MIL-C-5541.

4. CABLES SHOULD BE REARRANGED, IF PRACTICABLE, AT PLACES WHERE BENDING OCCURS TO FACILITATE TRUNK INSTALLATION.

5. FABRICATION DETAILS ARE TYPICAL DIMENSIONS AND MAY BE MODIFIED AS REQUIRED.

6. SEE FIGURE 3 FOR THE GROUNDING AND SHIELDING REQUIREMENTS FOR CABLES THAT EXIT THE WIREWAY TRUNK.

Figure 4-33(a) Mast Cables Located Within Wireway Trunk

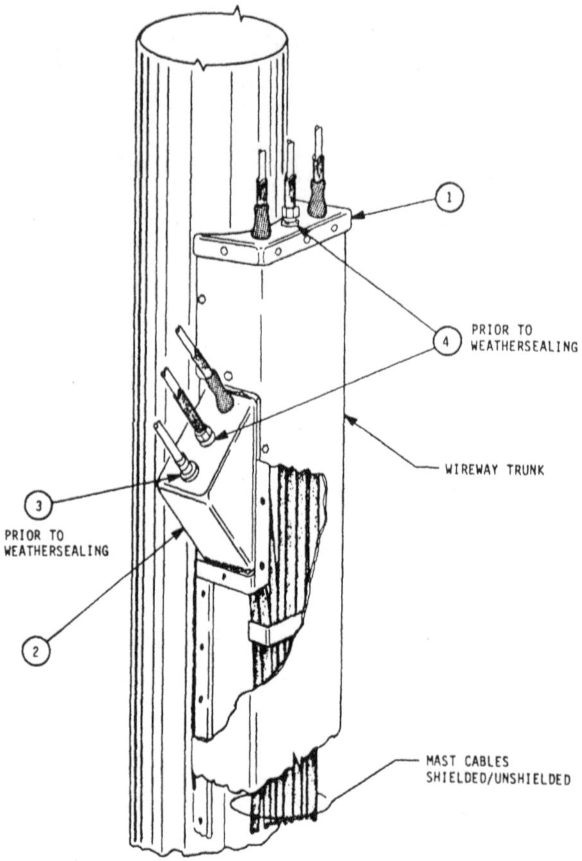

LIST OF MATERIAL			
ITEM NO.	PART	SPECIFICATION	NOTE
1	END CAP		1
2	BOX, BREAKOUT		1
3	ADAPTER, GROUNDING	COMMERCIAL	4
4	FITTING, CONDUIT	COMMERCIAL	4

NOTES:

1. FABRICATE END CAP AND BFEAK-OUT BOXES, AS REQUIRED, TO ACCOMODATE THE NUMBER OF AND TYPE OF CABLES THAT EXIT THE WIREWAY TRUNK. HOLES SHALL BE SIZED TO FIT THE REQUIRED GROUNDING ADAPTERS AND CONDUIT TERMINATION FITTINGS.

2. AFTER CABLE TYPES AND GROUNDING AND SHIELDING REQUIREMENTS HAVE BEEN PREDETERMINED, INSTALL THE CABLES THROUGH THE END CAPS, BREAKOUT BOXES, GROUNDING ADAPTERS AND THE CONDUIT FITTINGS.

3. MEASURE AND CUT THE REQUIRED LENGTHS OF SHIELDING CONDUIT TO SHIELD ALL REQUIRED CABLES.

4. INSTALL GROUNDING ADAPTERS AND FLEXIBLE SHIELDING CONDUIT. WEATHERSEAL AS SPECIFIED FOR CORROSION PROTECTION.

Figure 4-33(b) Mast Cables Located Within Wireway Trunk

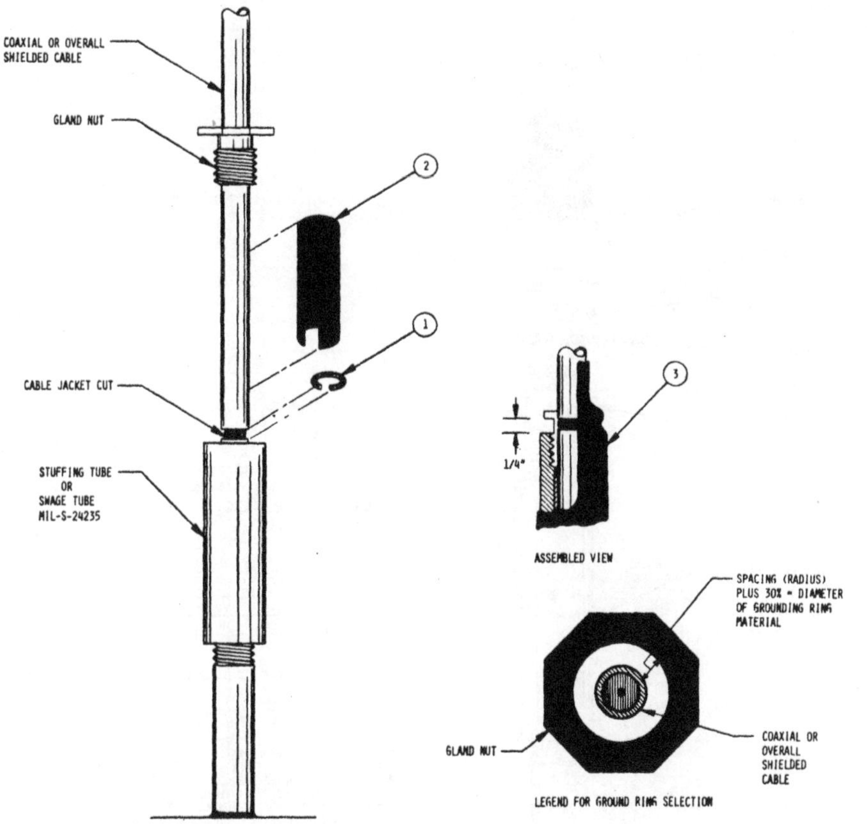

METHOD 1

LIST OF MATERIAL			
ITEM NO.	PART	SPECIFICATION	NOTE
1	GROUNDING RING	COMMERCIAL	4, 5, 8
2	COMPRESSION SLEEVE, SHIM STOCK	COMMERCIAL	7

NOTES:

1. THIS METHOD OF CABLE SHIELD GROUNDING APPLIES TO NEW CABLE INSTALLATIONS AND TO EXISTING INSTALLATIONS WHERE THE CABLE CANNOT BE REMOVED. THIS METHOD OF CABLE SHIELD GROUNDING IS PREFERRED DUE TO LOWER COST AND SIMPLICITY OF INSTALLATION. GROUNDING EFFECTIVENESS IS APPROXIMATELY THE SAME.

2. UNSCREW PACKING GLAND NUT FROM THE STUFFING TUBE AND MOVE IT SEVERAL INCHES UP THE CABLE AND TAPE. THE CABLE IS NOT REQUIRED TO BE REMOVED FROM THE CABLE HANGERS OR TO BE MOVED IN ANY WAY.

3. WITH A POCKET KNIFE OR SIMILAR TOOL, MAKE TWO CIRCULAR CUTS IN THE CABLE JACKET, ONE APPROXIMATELY FLUSH WITH THE TOP OF THE STUFFING TUBE OR SWAGE TUBE AND ANOTHER APPROXIMATELY ONE-FOURTH INCH HIGHER. ENSURE INSIDE OF GLAND NUT IS CLEAN AND FREE OF PAINT, SEALING COMPOUNDS, OR CORROSION. CLEANING WITH FINE SANDPAPER MAY BE REQUIRED.

4. REMOVE THE CUT SECTION OF THE CABLE JACKET. SELECT PROPER GROUNDING RING MATERIAL IN ACCORDANCE WITH THE LEGEND. CUT GROUNDING RING MATERIAL TO PROPERLY FIT (BUTT END-TO-END) THE CABLE AREA WHERE JACKET WAS REMOVED. COAT THE GROUNDING RING WITH ANTI-SEIZE COMPOUND OF MIL-T-22361. APPLY A COATING OF ANTI-SEIZE COMPOUND TO THE EXPOSED CABLE SHIELD AND TO THE THREADS OF THE GLAND NUT.

5. PLACE GROUNDING RING AROUND CABLE IN AREA WHERE JACKET WAS REMOVED. PLACE COMPRESSION SLEEVE AROUND CABLE JACKET AND GROUNDING RING. HOLDING COMPRESSION SLEEVE TIGHTLY AROUND CABLE AND GROUNDING RING, SLIDE GLAND NUT DOWN OVER SLEEVE AND THREAD INTO STUFFING TUBE. AFTER THREADS HAVE ENGAGED AT LEAST ONE-FOURTH INCH, REMOVE THE COMPRESSION SLEEVE AND COMPLETE TIGHTENING OF THE GLAND NUT AS REQUIRED FOR PACKING. WHEN COMPLETED, THE GROUNDING RING SHOULD BE LOCATED APPROXIMATELY MIDWAY ON THE INSIDE OF THE GLAND NUT WITH AN APPROXIMATE ONE-FOURTH INCH GAP BETWEEN THE TOP OF THE STUFFING TUBE AND THE BOTTOM OF THE HEX ON THE GLAND NUT. ADDITIONAL PACKING OR GLAND WASHERS MAY BE REQUIRED.

6. WEATHERSEAL AS SPECIFIED FOR CORROSION PROTECTION.

7. THE COMPRESSION SLEEVE IS USED ONLY TO COMPRESS THE GROUNDING RING WHILE REINSTALLING THE GLAND NUT. IT CAN BE CUT FROM APPROXIMATELY 0.005" MATERIAL.

8. ENSURE ELECTRICAL CONTACT BETWEEN GROUNDING RING AND CABLE SHIELD ON CORRUGATED OUTER SHIELD CABLES.

9. WHERE METAL PIPES AND TUBING ARE ROUTED THROUGH STUFFING TUBES, THE METHOD SHOWN HERE FOR CABLE SHIELD GROUNDING SHALL ALSO BE USED FOR PIPES AND TUBING GROUNDING.

10. GROUNDING RING SHALL BE ROUND CROSS SECTION, NEOPRENE SPONGE, FERREX WIRE, EMI MESH STRIP, CHROMERICS, INC., CARSON, CA, OR EQUAL.

Figure 4-34(a) Cable Shield Grounding Methods

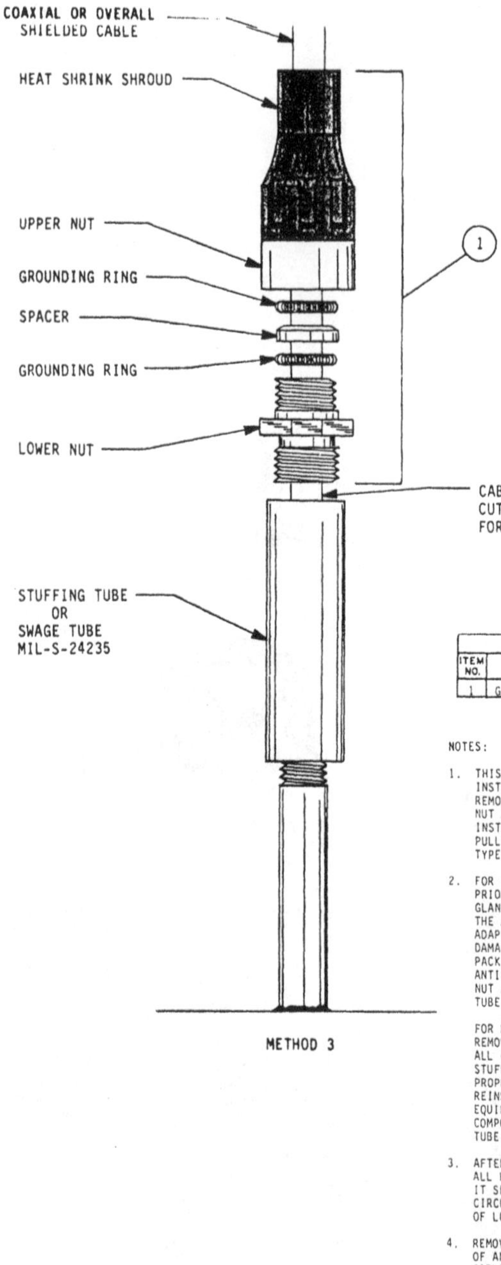

LIST OF MATERIAL			
ITEM NO.	PART	SPECIFICATION	NOTE
1	GROUNDING ADAPTER	COMMERCIAL	1

NOTES:

1. THIS METHOD OF CABLE SHIELD GROUNDING APPLIES TO NEW INSTALLATIONS AND TO EXISTING CABLES THAT CAN BE REMOVED AND ROUTED THROUGH THE GROUNDING ADAPTER. UPPER NUT AND LOWER NUT MUST BE LOOSELY THREADED TOGETHER WHEN INSTALLING THE ADAPTER ON CABLE TO ALLOW CABLE TO BE PULLED THROUGH ADAPTER. GROUNDING ADAPTERS SHALL BE TYPE CSGA, SIGMAFORM CORP., SANTA CLARA, CA, OR EQUAL.

2. FOR NEW CABLE INSTALLATIONS -
   PRIOR TO PULLING CABLE THROUGH THE STUFFING TUBE, REMOVE GLAND NUT AND REPLACE WITH THE ADAPTER, LOOSELY THREADING THE ADAPTER INTO TOP OF TUBE. PULL CABLE THROUGH ADAPTER AND STUFFING TUBE MAKING SURE CABLE DOES NOT DAMAGE ADAPTER COMPONENTS. INSTALL CABLE IN HANGERS. PACK STUFFING TUBE, COAT THREADS OF LOWER NUT WITH ANTI-SEIZE COMPOUND OF MIL-T-22361, AND TIGHTEN LOWER NUT AS REQUIRED FOR PACKING. ADAPTER CHOSEN MUST MATCH TUBE SIZE (A,B,C,ETC.).

   FOR RETROFIT INSTALLATIONS -
   REMOVE CABLE FROM TERMINATING EQUIPMENT AND REMOVE FROM ALL CABLE HANGERS DOWN TO TOP OF STUFFING TUBE. REMOVE STUFFING TUBE GLAND NUT AND SLIDE OFF OF CABLE. CHOOSE PROPER SIZE GROUNDING ADAPTER AND SLIDE DOWN OVER CABLE. REINSTALL CABLE IN HANGERS AND RECONNECT TO TERMINATING EQUIPMENT. COAT THREADS OF LOWER NUT WITH ANTI-SEIZE COMPOUND OF MIL-T-22361. THREAD ADAPTER INTO STUFFING TUBE AND TIGHTEN LOWER NUT AS REQUIRED FOR PACKING.

3. AFTER CABLE HAS BEEN PERMANENTLY INSTALLED IN PLACE AND ALL HANGERS ARE TIGHTENED, UNSCREW UPPER NUT AND MOVE IT SEVERAL INCHES UP THE CABLE AND TAPE. MAKE TWO CIRCULAR CUTS IN CABLE JACKET, ONE FLUSH WITH THE TOP OF LOWER NUT AND ANOTHER ONE-HALF INCH UP.

4. REMOVE CUT SECTION OF CABLE JACKET AND APPLY A COATING OF ANTI-SEIZE COMPOUND OF MIL-T-22361 TO THE EXPOSED CABLE SHIELD, TO THE GROUNDING RINGS, AND TO ALL ADAPTER THREADS. LOWER THE UPPER NUT AND HAND-TIGHTEN, MAKING SURE THE GROUNDING RINGS FALL INTO THE SLOT CUT IN THE CABLE JACKET.

5. WEATHER SEAL BY APPLYING HEAT TO TOP OF WEATHER SHROUD AS REQUIRED FOR PROPER SHRINKAGE AROUND CABLE. AFTER INSTALLATION, PERIODIC TIGHTENING OF THE ADAPTER FOR CABLE WEATHER SEALING IS NOT REQUIRED.

Figure 4-34(b) Cable Shield Grounding Methods

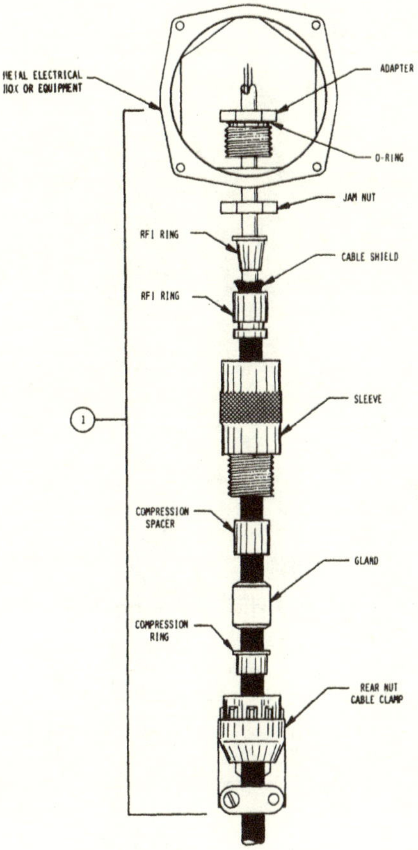

METHOD 6

LIST OF MATERIAL			
ITEM NO.	PART	SPECIFICATION	NOTE
1	ADAPTER, CABLE SHIELD GROUNDING	COMMERCIAL	4

NOTES:

1. THIS METHOD OF CABLE SHIELD GROUNDING APPLIES TO NEW AND RETROFIT INSTALLATIONS AND APPLIES ONLY TO CABLES WITH AN OVERALL SHIELD.

2. PRIOR TO ADAPTER INSTALLATION, ALL THREADED PARTS OF THE ADAPTER AND THE AREA WHERE THE ADAPTER JAM NUT CONTACTS THE BOX OR EQUIPMENT SHALL BE CLEANED AND COATED WITH ANTISEIZE COMPOUND OF MIL-T-22361.

3. INSTALL ADAPTER IN BOX OR EQUIPMENT AND SECURE WITH JAM NUT. PREPARE END OF CABLE ENSURING CABLE SHIELD IS PROPERLY CUT TO FIT RFI RINGS. ASSEMBLE ADAPTER ON CABLE AS SHOWN. TIGHTEN REAR NUT SECURELY FOR WEATHER SEALING.

4. CABLE SHIELD GROUNDING ADAPTERS SHALL BE ONE OF THE FOLLOWING, OR EQUAL:

   SUNBANK ELECTRONICS, INC., PASO ROBLES, CA, SERIES SE 93, STYLE A, TYPE 3, FINISH 34 (TYPE SHOWN ON THIS FIGURE)

   GLENAIR, INC., GLENDALE, CA, SERIES G1218, TYPE B, FINISH NF

   ELECTRO-ADAPTER, INC., CHATSWORTH, CA, SERIES EA90, STYLE 1, TYPE 1, FINISH 555

5. THE METHOD SHOWN FOR CABLE SHIELD GROUNDING DEPICTS ONLY ONE ACCEPTABLE METHOD. OTHER METHODS MAY ALSO BE ACCEPTABLE SUBJECT TO APPROVAL.

Figure 4-34(c) Cable Shield Grounding Methods

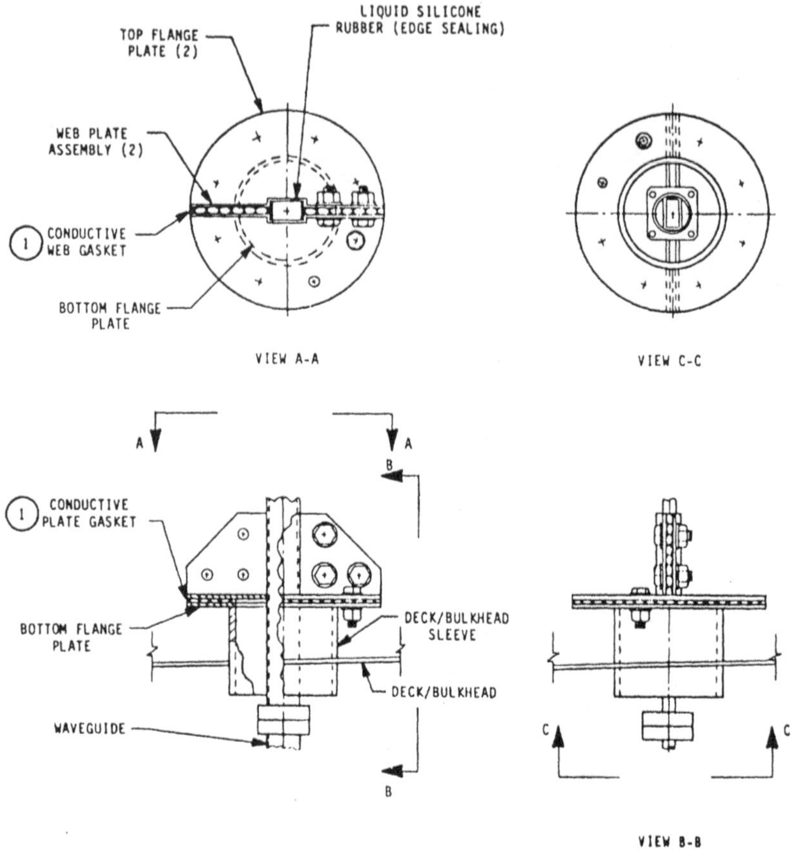

LIST OF MATERIAL			
ITEM NO.	PART	SPECIFICATION	NOTE
1	GASKET MATERIAL, CONDUCTIVE	COMMERCIAL	1

NOTES:

1. CONDUCTIVE GASKET MATERIAL SHALL BE CHROMERICS, INC., CARSON, CA, "POLASHEET", OR EQUAL. THE MOUNTING SURFACES FOR THE GASKET SHALL BE CLEANED TO BRIGHT METAL AND COATED WITH ANTI-SEIZE COMPOUND OF MIL-T-22361 PRIOR TO INSTALLING GASKET.

2. SPLIT SLEEVE INSTALLATION SHOWN IS AS DETAILED IN EIMB SERIES, NAVSEA 0967-000-0110.

Figure 4-35 Waveguide Grounding

transmitter spaces for connection of each radio transmitter cabinet enclosure. Similarly, antenna tuners and couplers are grounded to the keel ground plates, or, to keep cable lengths short, to the transmitter enclosures. For all other equipment and items, a size 1 AWG cable, connected to the ground plates or transmitter enclosures, is used as the main ground cable to which size 7 AWG branch ground cables are attached, as illustrated in Figure 4-36. By use of the

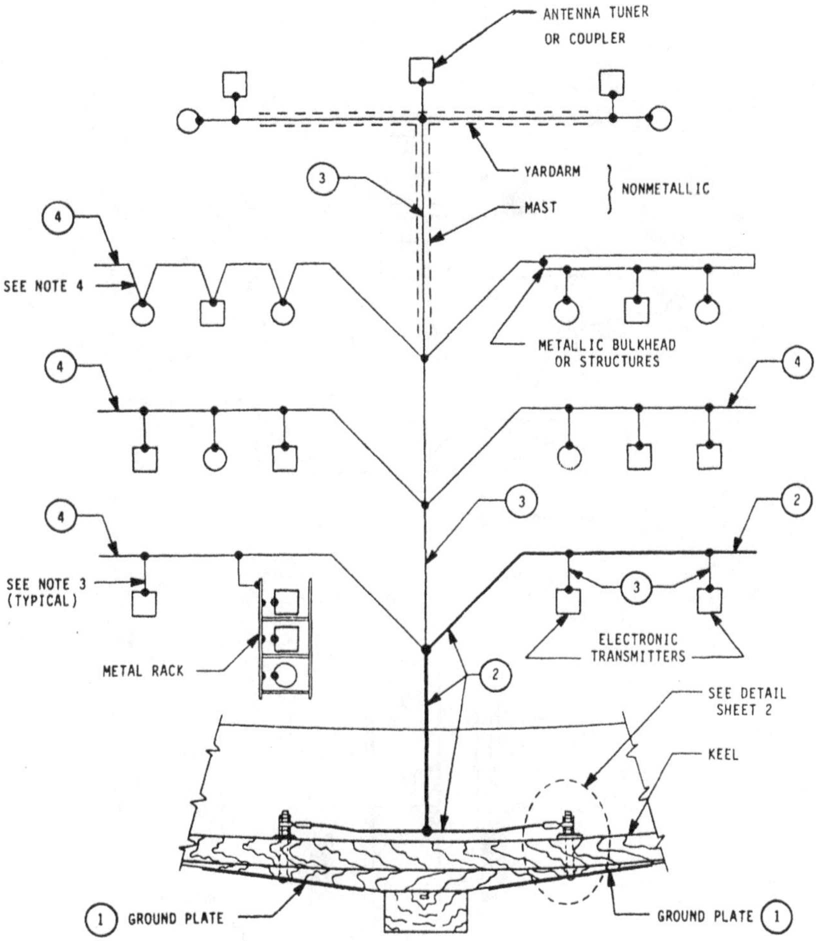

LIST OF MATERIAL			
ITEM NO.	PART	SPECIFICATION	NOTE
1	PLATE, COPPER, GROUNDING	QQ-C-576	1
2	CABLE, COPPER, STRANDED, NO. 1/0 AWG	MIL-C-24643	2,3,4
3	CABLE, COPPER, STRANDED, NO. 1 AWG	MIL-C-24643	2,3,4
4	WIRE, COPPER STRANDED, NO. 10 AWG	MIL-C-24643	2,3,4

SYMBOL LEGEND:

□ - ELECTRONIC EQUIPMENT

○ - ELECTRICAL EQUIPMENT OR METAL ITEMS

NOTES:

1. GROUND PLATES SHALL BE LIGHT, COLD-ROLLED, OXYGEN-FREE COPPER, APPROXIMATELY ONE-EIGHTH INCH THICK AND SHALL PROVIDE APPROXIMATELY 16 SQUARE FEET OF TOTAL SURFACE AREA EACH SIDE OF THE KEEL.

2. ALL BRANCH GROUND CABLES NOT SPECIFICALLY IDENTIFIED AS TO SIZE SHALL BE NO. 7 AWG STRANDED COPPER CABLE.

3. IN ACTUAL INSTALLATIONS, BRANCH CABLES MAY CONNECT DIRECTLY TO EACH EQUIPMENT GROUND CONNECTION TERMINAL.

4. THE CABLE SIZES DETAILED HEREIN ARE SPECIFIED FOR FULL-SIZE SHIPS SUCH AS AN MSO OR AN MCM. SIZING FOR SMALLER SHIPS SHALL BE AS APPROPRIATE.

Figure 4-36(a) Ground System, Nonmetallic Hull Ships

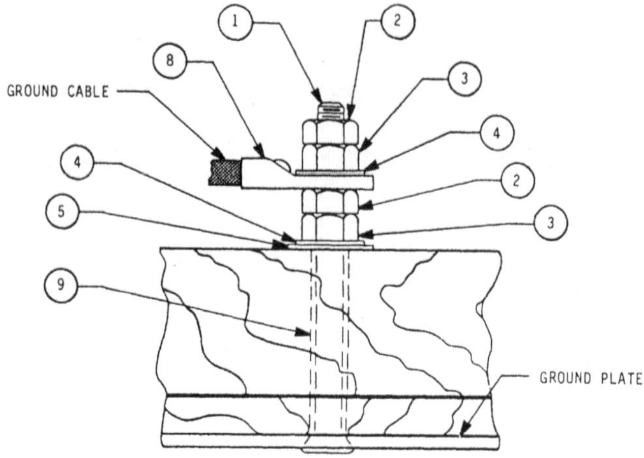

GROUND PLATE BOLT DETAILS (NOTES 1,2, AND 3)

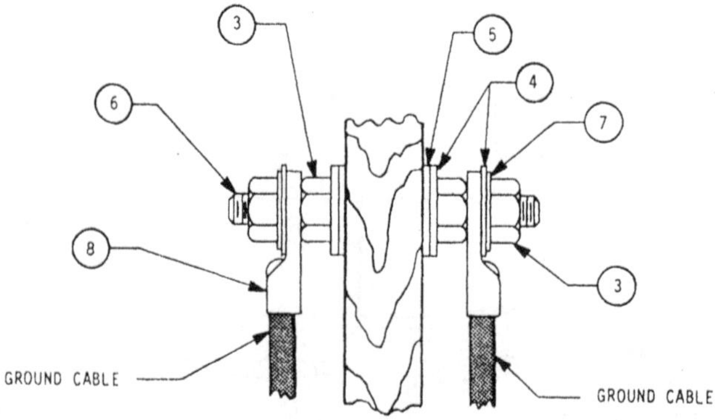

METHOD OF PASSING GROUND BUS THROUGH
WATERTIGHT BULKHEADS OR DECKS (NOTES 2 AND 3)

LIST OF MATERIAL			
ITEM NO.	PART	SPECIFICATION	NOTE
1	BOLT, COPPER		1,2
2	NUT, JAM, COPPER		
3	NUT, HEX, COPPER		
4	WASHER, COPPER		
5	WASHER, RUBBER		
6	STUD, COPPER		
7	WASHER, LOCK COPPER		
8	TERMINAL, LUG, COPPER		
9	SLEEVE, NONMETALLIC		3

NOTES:

1. HEAD OF THE GROUNDING PLATE THROUGH-BOLT SHALL BE BRAZED TO THE COPPER GROUNDING PLATE.

2. SIZE OF THE GROUNDING PLATE THROUGH-BOLT AND THROUGH-STUD SHALL AT LEAST EQUAL THE SIZE OF THE ASSOCIATED CABLE.

3. PROTECTION SHALL BE PROVIDED FOR THE STUD AGAINST THE CORROSIVE EFFECTS OF DAMP WOOD. THIS PROTECTION SHALL BE BY A NONMETALLIC SLEEVE.

Figure 4-36(b) Ground System, Nonmetallic Hull Ships

branch cables, all equipment using electrical power and all fuel tanks, water tanks, engines, engine control apparatus, metal screens, ducts, and metallic deck items such as standing rigging, cranes, king posts, liferails, and ladders are connected to the ship ground system.

4-2.2.4 RF Bonding Procedures for EMI Control

To preclude the formation of nonlinear junctions (and thereby reduce the potential for hull-generated intermodulation interference), ship topsides should be kept as free as possible of all pinned, snap-linked, and chain-linked metallic discontinuities. All metal-to-metal joints must be class-A-bonded, unless required to be removable. Further, the mating of dissimilar metals by bolting or riveting must be minimized. The joining of aluminum to steel should be accomplished by welding using bimetallic bonded joints. Loose metallic items such as pipes, cables, tools, and portable rigging should not be stowed topside except where absolutely necessary (as in the case of anchor chains).

In an attempt to minimize the possibility of nonlinear junction intermodulation sources, all ships having six or more HF transmitters must apply the following control measures. (Ships with less than six transmitters are to suppress only those sources positively identified through onboard EMI testing.)

a. Metallic walking ropes and hand safety ropes are not to be used on yardarms. Instead, nonmetallic rails or all-welded rails are to be used.
b. Rigging such as halyard downhauls, full dress rigging, awning lines, lifeboat lines, and other similar lines are to be nonmetallic. Metallic standing rigging must be bonded to ground as shown in Figure 4-37.
c. Aluminum or all-welded steel liferails are to be used at all deck edge areas not requiring personnel access or clear deck. Where clear deck edge is required, Kevlar nonmetallic lifelines must be installed. (Kevlar is a registered trademark of E. I. Dupont DeNemours and Co., Inc.) Additionally, access openings less than six feet wide must be protected with nonmetallic rope.
d. Life and safety nets and net frames, where determined to be a source of intermodulation noise, must be fabricated from nonmetallic material (except in heat or blast areas) or bonded as shown in Figure 4-38.
e. Portable flagstaffs, jackstaffs, and stanchions, where determined to be a source of intermodulation noise, must be either fabricated from nonmetallic material or bonded as shown in Figure 4-39.
f. Metallic awning rigging must be disassembled and stowed when the ship is under way, and awning stanchions, braces, and spreaders must be nonmetallic.

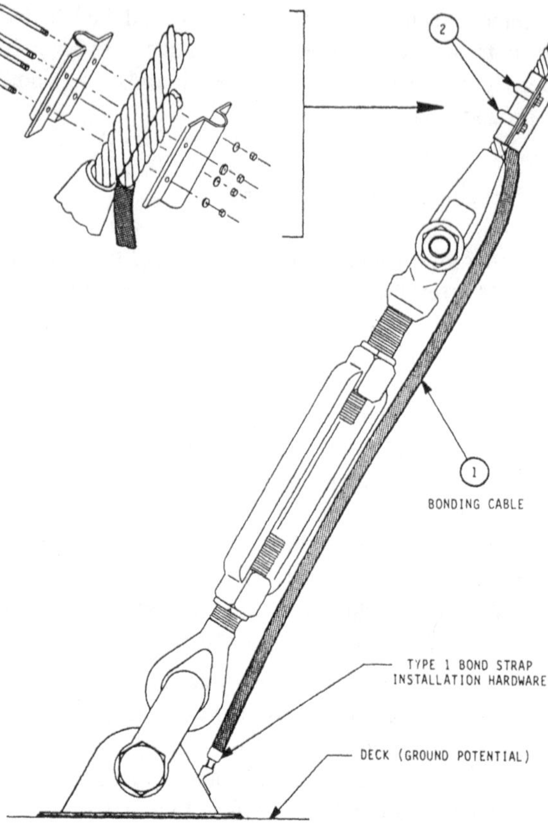

LIST OF MATERIAL			
ITEM NO.	PART	SPECIFICATION	NOTE
1	CABLE, WELDING, TYPE TRXF 84	MIL-C-915/21	1
2	U-BOLT ASSEMBLY		1

NOTES:

1. BONDING CABLE SHALL BE MEASURED AND CUT TO PROPER LENGTH. ONE END SHALL BE EQUIPPED WITH A LUG TERMINAL INSTALLED THE SAME AS A TYPE I BOND STRAP. THE UPPER END SHALL BE ATTACHED TO THE WIRE-ROPE STAY BY CLEANING BOTH CABLES AT POINT OF CONTACT AND APPLYING MIL-T-22361 ANTI-SEIZE COMPOUND THEN CLAMPING THE CABLES BY THE METHOD SHOWN. OVERALL WEATHERSEALING SHALL BE PROVIDED AS SPECIFIED THEREIN.

Figure 4-37 Standing Rigging, Bonding

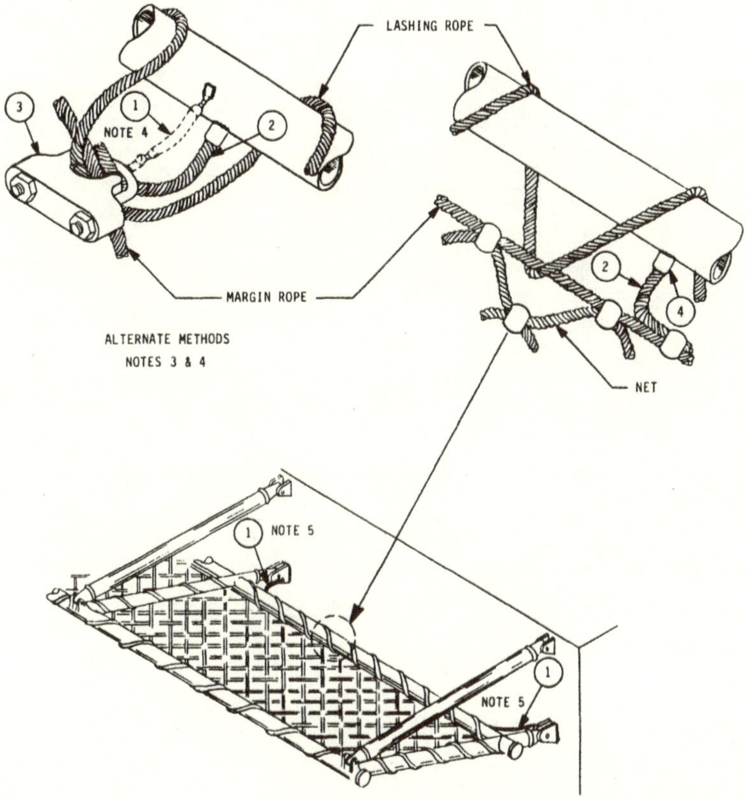

LIST OF MATERIAL			
ITEM NO.	PART	SPECIFICATION	NOTE
1	BOND STRAP, TYPE I		1
2	BONDING CABLE, CRES		2,3
3	U-BOLT, CRES		3,4
4	SWAGE SLEEVE		2

NOTES:

1. A TYPE I BOND STRAP SHALL BE INSTALLED ACROSS EACH NET FRAME HINGE. THE TYPE I BOND STRAP MAY BE INCREASED IN LENGTH, IF NEEDED, TO ALLOW NETS TO RAISE AND LOWER.
2. SWAGE SLEEVE (OR SIMILAR DEVICE) SHALL BE CRIMPED TO BONDING CABLE AND WELDED TO NEW FRAME.
3. AS AN ALTERNATE METHOD, A CRES U-BOLT MAY BE INSTALLED AROUND THE MARGIN ROPE, LASHING ROPE, AND BONDING CABLE.
4. AS A SECOND ALTERNATE, THE LUG OF A TYPE I BOND STRAP MAY BE WELDED TO A U-BOLT AND INSTALLED AS SHOWN.
5. WHERE NETS ARE REQUIRED TO BE REMOVED PERIODICALLY FOR MAINTENANCE, A TYPE II BOND STRAP MAY BE INSTALLED.

Figure 4-38 Metallic Life and Safety Nets, Bonding

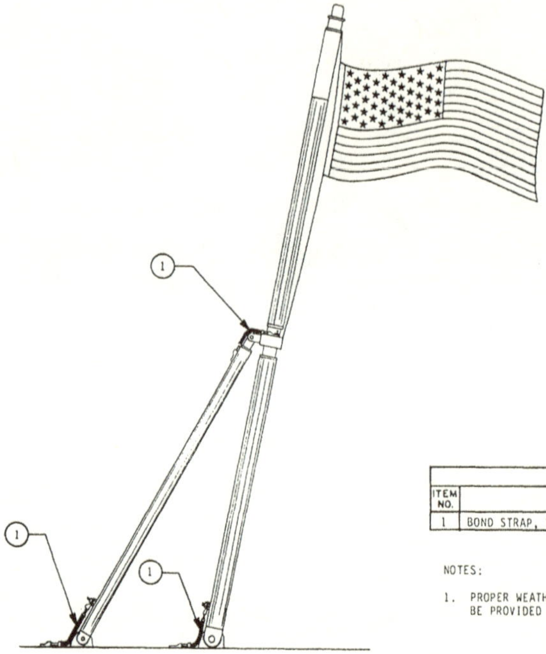

LIST OF MATERIAL			
ITEM NO.	PART	SPECIFICATION	NOTE
1	BOND STRAP, TYPE II		1

NOTES:

1. PROPER WEATHERSEALING OF BOND STRAP STUD TERMINAL SHALL BE PROVIDED AS SPECIFIED THEREIN.

Figure 4-39 Metallic Flagstaff or Jackstaff, Bonding

g. Metallic inclined ladders must be grounded as shown in Figure 4-40 or replaced with ladders made of nonmetallic material. Metallic vertical ladders are considered satisfactorily grounded when tightly secured bolts are used. Climber safety rails are considered satisfactorily grounded when installed with welded brackets. Brackets clamped to ladder rungs must use a type II bond strap at these points, with the welded end of the bond strap attached to the hull structure and the detachable end bolted to the safety rail.

h. Portable liferails are to be constructed of nonmetallic material except in heat or blast areas.

i. Armored cables must not be used for new design ships. On ships where armored cable already exists, the cable will be relocated inside the mast as shown in Figure 4-41 or within a wireway enclosure as depicted in Figure 4-33.

j. Expansion joints must be bonded as illustrated in Figure 4-42.

k. Tilting antenna platforms must be bonded as shown in Figure 4-43.

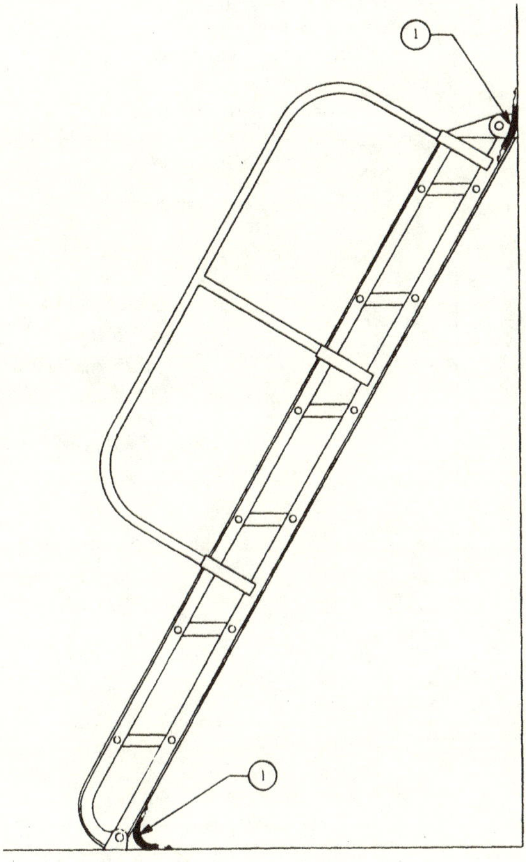

LIST OF MATERIAL			
ITEM NO.	PART	SPECIFICATION	NOTE
1	BOND STRAP, TYPE I OR II		1

NOTES:

1. INCLINED-TREAD LADDERS SHALL BE BONDED TO GROUND POTENTIAL BY THE INSTALLATION OF A BOND STRAP ACROSS ONE TOP AND ONE BOTTOM PINNED MOUNT. TYPE I BOND STRAPS ARE PREFERRED. WHERE THE LADDER MUST BE PERIODICALLY REMOVED, TYPE II BOND STRAPS SHALL BE INSTALLED. TYPE II BOND STRAPS SHALL BE WELDED TO SHIP HULL OR STRUCTURE AND BOLTED TO THE LADDER. PROPER WEATHERSEALING OF THE BOND STRAP STUD TERMINAL SHALL BE PROVIDED.

Figure 4-40 Metallic Inclined-Tread Ladders, Bonding

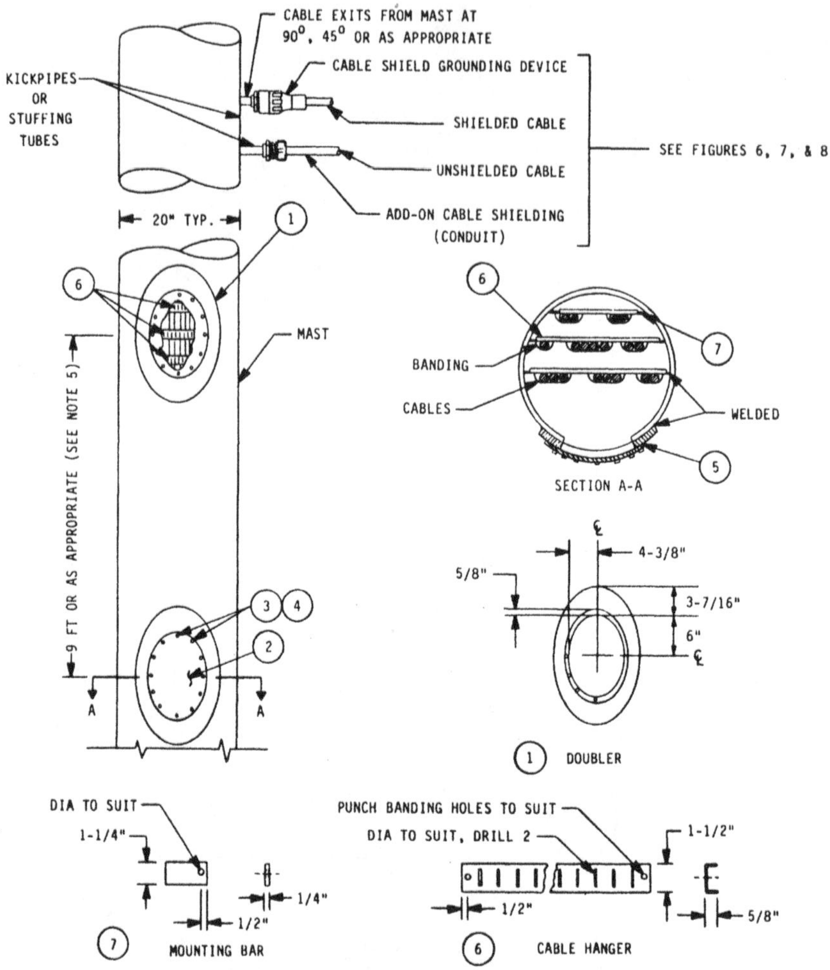

LIST OF MATERIAL			
ITEM NO.	PART	SPECIFICATION	NOTE
1	DOUBLER, MED STL, 40.8# PL.	MIL-S-22698	1,2
2	COVER, MED STL, 10.2# PL.	MIL-S-22698	1,2
3	BOLT, MACH, STL-ZINC PLATED		
	3/8"-16 UNC 2A HEX HD	MIL-S-1222	1,2
4	WASHER, STL-ZINC PLATED SPLIT		
	LOCK, 3/8"		1,2
5	GASKET, RUBBER, 1/8" THK	MIL-R-900	1
6	CABLE HANGER		3
7	MOUNTING BAR		4

NOTES:

1. FABRICATION DETAILS ARE TYPICAL DIMENSIONS AND MAY BE MODIFIED TO SUIT OTHER SHIPS AS REQUIRED.
2. FOR ALUMINUM MASTS, COVERS AND DOUBLER PLATES SHALL BE FABRICATED FROM ALUMINUM MATERIAL.
3. NUMBER AND SPACING OF CABLE HANGERS WILL BE DETERMINED BY CABLE REQUIREMENTS AND MAST SIZE.
4. MOUNTING BARS SHALL BE INSTALLED BY WELDING TO INSIDE OF MAST.
5. EXTERNAL ACCESS HOLES ARE NOT REQUIRED IN MASTS THAT ARE LARGE ENOUGH TO PERMIT INTERNAL CABLE INSTALLATIONS AND MAINTENANCE.

Figure 4-41 Mast Cables Located Within Mast, Typical

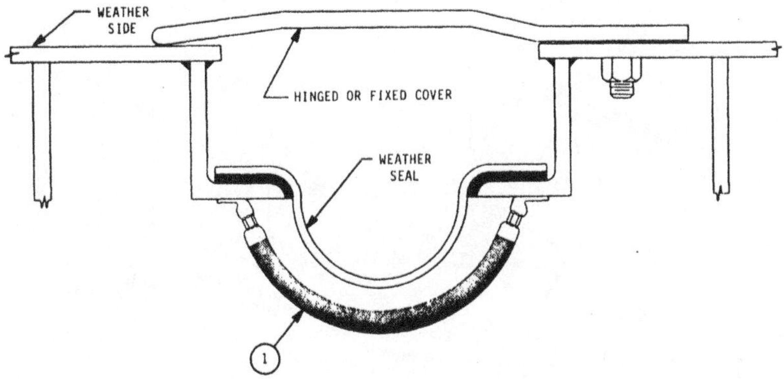

LIST OF MATERIAL			
ITEM NO.	PART	SPECIFICATION	NOTE
1	BOND STRAP, TYPE I		1,2

NOTES:

1. BOND STRAPS SHALL BE INSTALLED ACROSS EXPANSION JOINTS AT INTERVALS OF APPROXIMATELY 5 FEET AND LOCATED ON THE SIDE OF THE JOINT NOT EXPOSED TO THE WEATHER.
2. THE LENGTH OF THE BOND STRAP SHALL BE SUFFICIENT TO PERMIT MAXIMUM EXCURSION OF THE EXPANSION JOINT.

Figure 4-42 Expansion Joints, Bonding

l. Large or long portable metallic items or equipment such as fog nozzles, davits, and personnel stretchers stowed within 50 feet of an HF antenna must be insulated from contact with the ship hull structure by insulated hangers, clips, or brackets. Insulating material may be weather-resistant, heat-shrinkable tape or tubing, rubber matting, plastics, epoxy, fiberglass, or other similar materials.

m. Masts, mast braces, king posts, and similar deck structures bolted in place must be grounded by type I bond straps spaced equally around each structure as seen in Figure 4-44.

Care in preparing the surface for good bonding is very important. Surface preparation for installation of welded or brazed bond straps (type I and type II) and welded studs must be accomplished by cleaning to bare metal the areas where bond strap lugs are to be welded, brazed, or bolted. Cleaned areas and all threaded hardware must be coated with an antiseize compound prior to installation of bolted bond straps.

Bond strap installation hardware such as nuts, bolts, washers, and studs are to be either $\frac{5}{16}$-inch or ⅜-inch as appropriate. For topside areas, mounting hardware must be corrosion-resistant steel except where aluminum studs are required. In areas other than topside, the mounting hardware (except studs) must

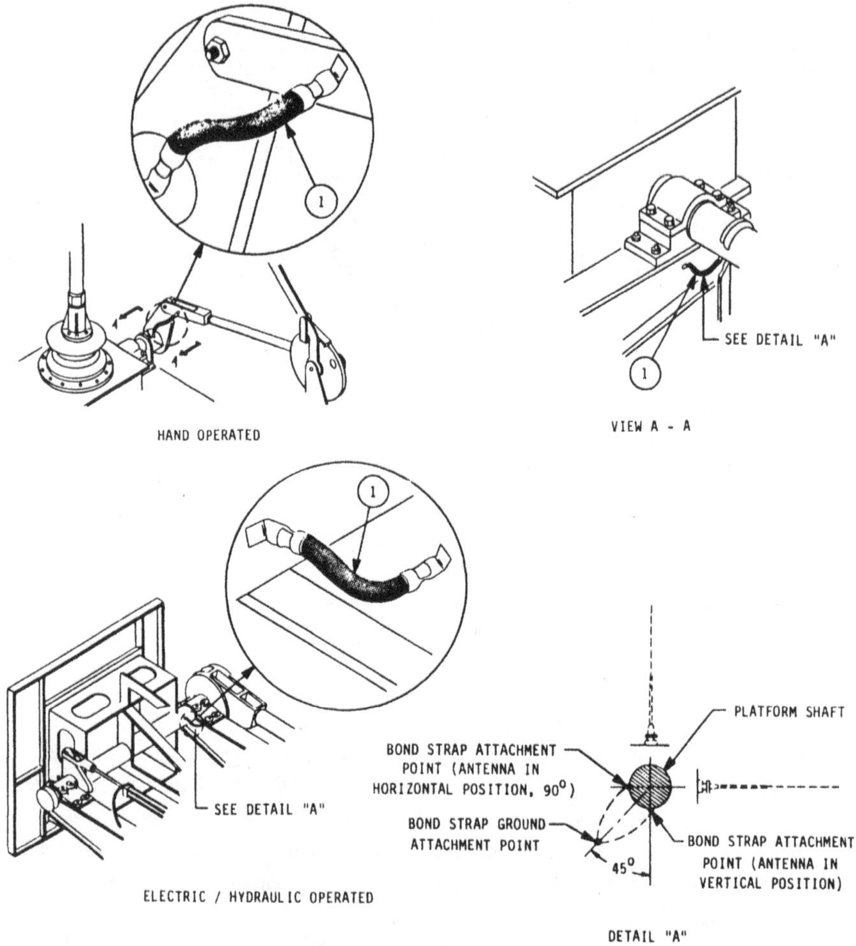

LIST OF MATERIAL			
ITEM NO.	PART	SPECIFICATION	NOTE
1	BOND STRAP, TYPE 1		1

NOTES:

1. BOND STRAP LENGTH AND METHOD OF INSTALLATION SHALL ALLOW FOR MAXIMUM TRAVEL OF ANTENNA TILTING MECHANISM.

Figure 4-43 Tilting Antenna Mounts, Bonding

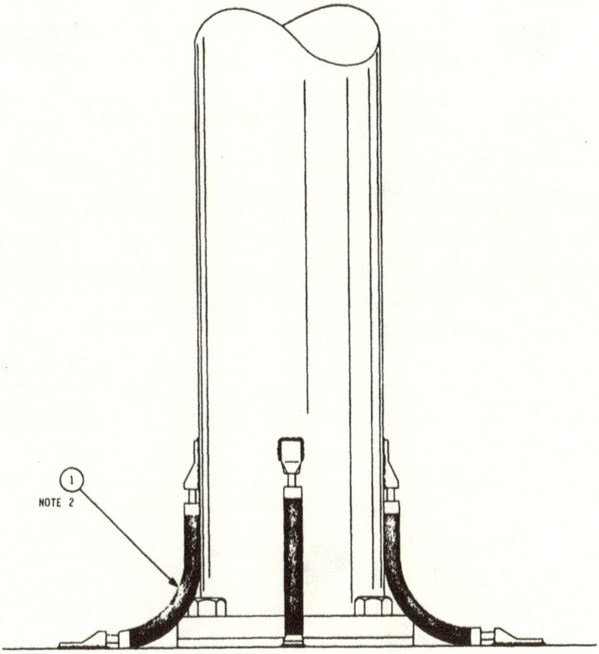

LIST OF MATERIAL			
ITEM NO.	PART	SPECIFICATION	NOTE
1	BOND STRAP, TYPE 1		1,2

NOTES:

1. BOND STRAPS SHALL ONLY BE INSTALLED ON MASTS WHICH ARE BOLTED IN PLACE.
2. BOND STRAPS SHALL BE INSTALLED IN ACCORDANCE WITH THE FOLLOWING:

MAST DIAMETER	NO. OF BOND STRAPS
20 INCHES OR LARGER	4
20 INCHES TO 8 INCHES	2
LESS THAN 8 INCHES	1

BOND STRAPS SHALL BE EQUALLY SPACED AROUND MAST.

Figure 4-44 Metallic Masts, Bonding

be plated steel. Studs in other than topside areas may be either aluminum or plated steel as appropriate. Methods of attaching unwelded bond straps are shown in Figure 4-45.

Bond straps are to be installed so as to permit immediate inspection and replacement, and mounted in such a manner that vibration, expansion, contraction, or relative movement will not break or loosen the strap connection. In-

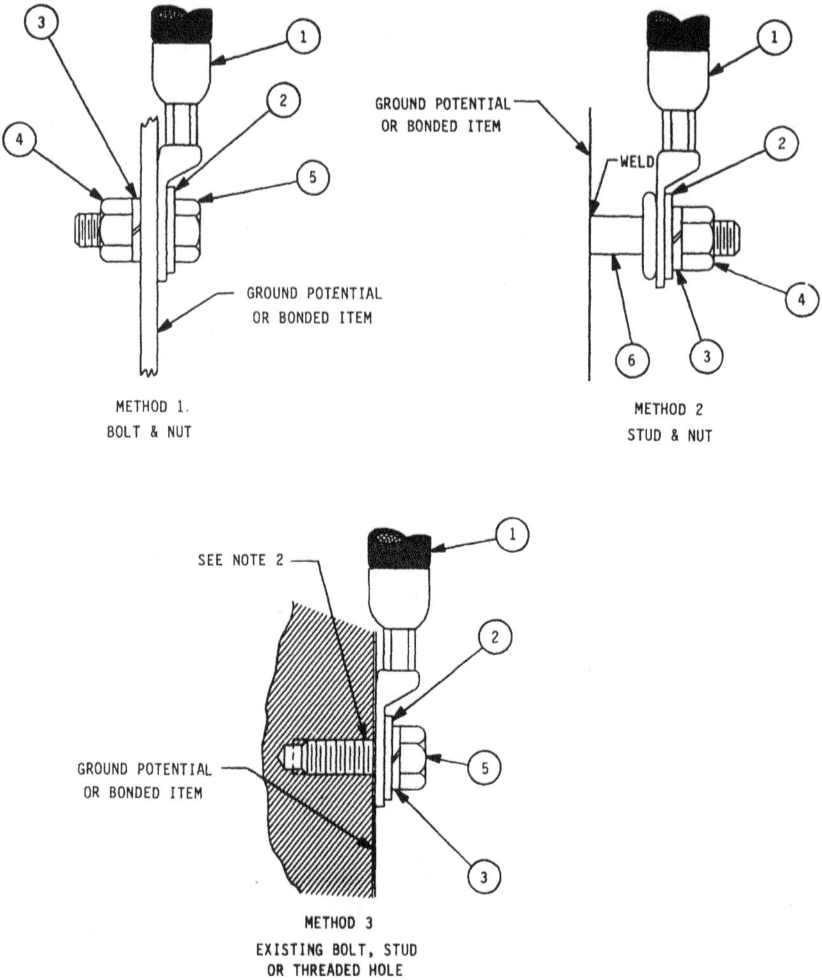

LIST OF MATERIAL			
ITEM NO.	PART	SPECIFICATION	NOTE
1	BOND STRAP, TYPE II, III OR IV		1,2,
2	WASHER, FLAT	FF-W-92	5
3	LOCKWASHER, SPLIT	FF-W-84	5
4	NUT	MS35425,and	
		FF-N-836	5
5	BOLT	MIL-S-1222	5
6	STUD, SHOULDER OR COLLAR	MIL-S-24149	3

NOTES:

1. EXISTING BOLTS, STUDS, OR THREADED HOLES MAY BE USED FOR BOND STRAP INSTALLATION.

2. THE INSTALLATION PROCEDURES FOR BOLTED BOND STRAPS SHALL PROVIDE FOR A CLEAN METAL-TO-METAL CONTACT BETWEEN THE BOND STRAP AND THE MATING SURFACE.

3. STUDS USED FOR BOND STRAP ATTACHMENT SHALL BE A COLLAR TYPE. TO PERMIT WELDING, STUDS SHALL CORRESPOND TO THE MATING SURFACE, ALUMINUM STUDS FOR ATTACHMENT TO ALUMINUM SURFACES AND STEEL STUDS FOR ATTACHMENT TO STEEL SURFACES. STUDS USED FOR TYPE II BOND STRAP INSTALLATIONS SHALL CONFORM TO THE FOLLOWING REQUIREMENTS OF MIL-S-24149:

STUD SIZE	- 3/8"-15
STEEL STUDS	-TYPE V, CLASS 4, CRES
ALUMINUM STUDS	_TYPE IV, CLASS 3

4. THREADED HARDWARE SHALL BE PREPARED AND SEALED IN ACCORDANCE WITH THE REQUIREMENTS OF 5.5.2, 5.5.3, AND 5.5.4.

5. FOR SHIPBOARD EXTERIOR APPLICATIONS, ITEMS 2, 3, 4, AND 5 SHALL BE CORROSION RESISTANT STEEL.

Figure 4-45 Methods of Attaching Nonwelded Bond Straps

stallation of bond straps must not interfere with the structural integrity of cabinets or enclosures, or weaken any item to which the strap is attached, or restrict the movement of hinged or movable items. Where convenient, existing bolts, studs, or threaded holes may be used for bond strap installation.

The lug ends of type I and type II bond straps which have been welded in place must be weather-sealed by priming and painting the lugs and welded areas. The cable jackets of these type bond straps do not require painting; painting the jackets, however, will not affect the bond strap performance.

Type II and type III bond straps installed on threaded studs or fastened by bolts must be weather-sealed by coating the lugs and associated hardware with MIL-S-81733 sealing compound. After installation, painted areas affected are to be restored to the original paint finish. Bond straps installed in areas other than topside do not require weather-sealing or painting.

Antiseize compounds used between metal surfaces to be bonded preserve grounding conductivity. These compounds maintain the quality of grounding by preventing oxidation or corrosion in the ground path. The compounds are used only in areas where metal-to-metal contact through the compound can be maintained under pressure such as with threaded bolting. After application of the antiseizing compound and attachment of the bond strap, the union must be sealed with MIL-S-45180 sealing compound to prevent the antiseize material from melting and running under high temperatures.

Examples of potential topside nonlinear junction intermodulation interference sources and RF bonding are shown in Figures 4-46 through 4-51.

4-2.3 Nonmetallic Topside Material Techniques

In the previous section it was noted that two primary means for the reduction of hull-generated intermodulation interference are: (1) to replace potential metallic noise contributors with nonmetallic items; and (2) to use insulation material for isolation of the offending source from the metallic hull. These methods are not new in concept. Improved materials and installation techniques have been implemented continually throughout the years, however, and are proving remarkably effective in lessening intermodulation and broadband noise.

Traditional naval lifelines made of metal are notorious generators of intermodulation interference. Long, relatively free of deck obstructions, clasped to vertical metal posts, these lifelines act as natural parasitic antenna elements to intercept RF energy such as HF transmissions prevalent aboard ship. The coupled energy is then conducted along the lines to terminal points of connectors and turnbuckles, making and breaking contact at the stanchion hooks. The resultant rapidly intermittent metallic contact creates arcing and intermodulation EMI. Furthermore, in addition to being generators of noise, lifelines in the field of view of microwave antennas perturb the radiation patterns; those in HF fields

(a)

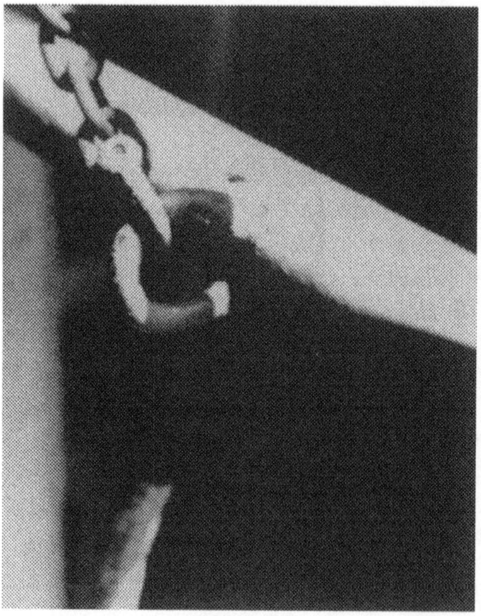

(b)

Figure 4-46 Typical Shipboard Unbonded Potential Noise Sources (Intermittent Metal Contact)

(a)

(b)

Figure 4-47 Typical Shipboard Unbonded Potential Noise Source (Intermittent Metal Contact)

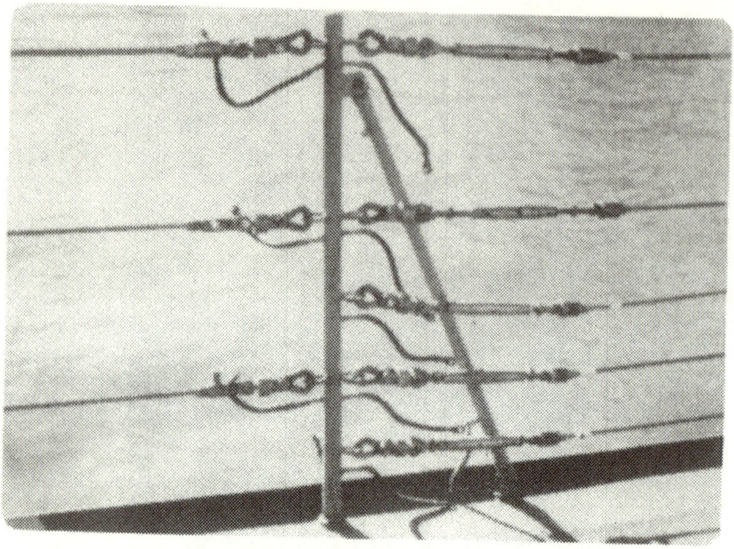

(a)

(b)

Figure 4-48 Bond Straps Across Lifeline Connections

(a)

(b)

Figure 4-49 Bonding of Rotatable Joints

(a)

(b)

Figure 4-50 Bonding of Ship Exhaust Stacks and Pipes

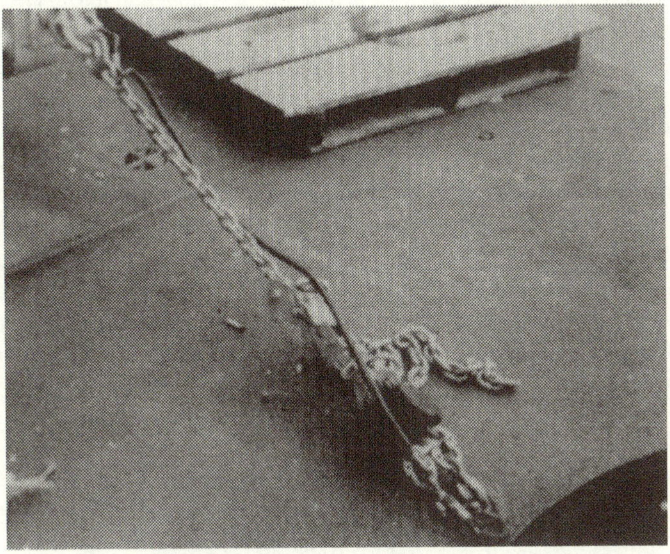

(a)

(b)

Figure 4-51 Ineffective Bond Strap

create RF burn hazards to personnel; and all metal connection hardware is subject to corrosion, thereby aggravating the production of intermodulation noise by nonlinear junctions. Because of these unfortuante tendencies shipboard lifelines have been a focal point of EMI engineering practices for many years.

Long-term application and evaluation of various material and design techniques for improving lifeline EMC have been continuous. The earliest experiments involved the use of prestretched, double-braided nylon, and, later, mylar-type nonmetallic ropes. These alternatives to metallic lifelines were deemed unsatisfactory, however, as the material stretched and sagged. Glass-reinforced plastic lines seemed to offer good promise, but these too proved inadequate after extended hardships of shipboard wear and tear. An example of a dangerously worn fiberglass lifeline that has suffered severe abrasion at a stanchion J-hook is pictured in Figure 4-52.

Figure 4-52 Worn Nonmetallic Lifeline

Contemporary plastics have been much improved in recent years in their resistance to the marine environment, in weight-to-strength ratio, and in their low-stretch characteristics. As a consequence, their servicability has been welcomed, and Kevlar lifelines currently are being employed as the standardized nonmetallic lifeline for US naval ships.[26]

Note, however, that emphasis is being given to reducing the use of lifelines and guardlines to a minimum aboard ship. Instead, fixed liferails of welded steel or aluminum are used wherever practicable. The present objective is to: (1) install welded liferails (i.e., having no hinged or moving connections) in all deck areas except those requiring removable stanchions such as at replenishment stations and safety nets; (2) use Kevlar nonmetallic lifelines where clear deck edges must be maintained; and (3) use polyester rope in place of metal chain for short guardlines. In such a manner all possible EMI sources normally created in shipboard liferails and lifelines are eliminated.

As means of reducing antenna radiation pattern disturbances, however, nonmetallic liferails and lifelines have not fared so well. Recent studies have concluded that, rather than being transparent to microwaves, nonmetallic rails such as those seen in Figure 4-53 interact with electromagnetic energy as much or more than do metal rails of equivalent size and form. In fact, indications are that, with nonmetallic obstructions, radar antenna sidelobes are enhanced while mainbeam levels are reduced.[27] For this reason current topside design practices recommend that microwave antennas be placed on a pedestal high enough to radiate clearly over railings, as shown in Figure 4-54.

Figure 4-53 Nonmetallic Lifelines in Antenna Field of View

Figure 4-54 Nonmetallic Lifelines Below Antenna Field of View

There are many other instances where, in addition to the special case of lifelines, nonmetallic materials are used to reduce the occurrence of shipboard intermodulation noise generation. For example, it is now commonplace to use nonmetallic guy wires, life nets, flag boxes, inclined ladders, stanchions, flagstaffs, jackstaffs, and utility boxes. Moreover, several ever-present topside items such as fog nozzles, booms, davits, personnel stretchers, and pipes are isolated from the metal hull by insulated cradles and brackets. Examples of insulated devices are shown in Figure 4-55.

4-2.4 EMI Filtering Techniques

Good design, maintenance, grounding, bonding, and shielding practices quite often are still not sufficient to prevent some forms of EMI from reaching and degrading the performance of shipboard electrical and electronic equipment. Of course these EMI control techniques should be diligently applied to reduce the potential sources of interference to a minimum. Yet, in spite of the above engineering procedures, conducted interference sometimes will find a way to gain entrance. It is in such cases that filter devices can help.

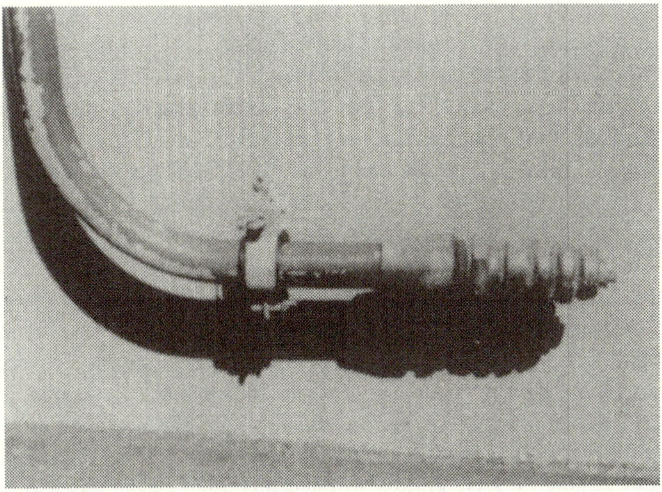

(a)

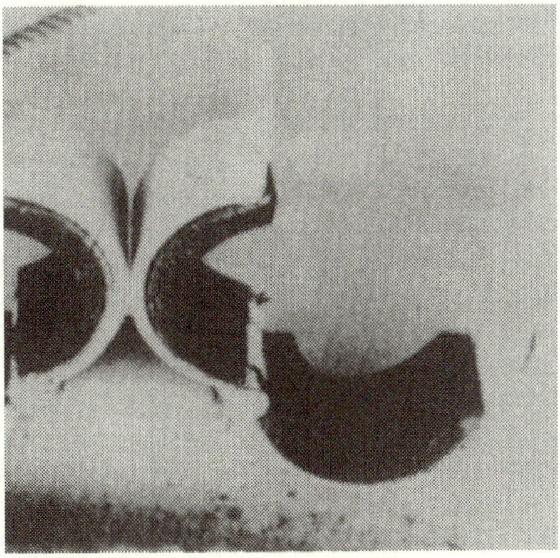

(b)

Figure 4-55 Insulated Brackets and Cradles

4-2.4.1 *Filter Classification and Characteristics*

Electronic filters are generally either (1) the reactance circuit type that employs discrete resistor, capacitor, and inductor (RLC) components specially configured to pass currents at certain frequencies and to block currents at other frequencies, or (2) the lossy line type that, rather than rerouting or reflecting unwanted signals, absorbs and dissipates them.

Reactance filters use series and parallel RLC combinations in familiar L, T, and pi networks. Using circuit resonance characteristics, these networks present a high impedance to interference flowing in the desired signal path while shunting the interference to ground via a very low impedance branch. In contrast, lossy filters are constructed of such materials as silver-coated ferrites that act strongly to attenuate undesired frequencies. In either case, the filters are so designed to discriminate against unwanted signals and to inhibit their conduction in the path of the desired signal. Filters are incorporated by manufacturers as an integral part of equipment and systems to achieve specified performance requirements; our main interest here, however, is after-the-fact filter applications to rid ship systems of EMI known to have disrupted or degraded mission operations.

Filters normally are classified as low pass, high pass, band pass, or band reject, depending on the intended method of excluding interference frequencies (as functionally illustrated in Figure 4-56). Low pass filters are most often used in EMC and are usually available in pi networks consisting of a series inductor and two capacitors in a three-branch circuit.[28]

Because they ordinarily are inserted in an active circuit in such a way that all circuit energy has to flow through some part, filters must accomplish their function without impairing normal operations. Ideally, there would be no adverse effect at all on the desired signal upon addition of the filter, but in practice, a small amount of signal attenuation does occur. Therefore, one important measure of a filter's quality is its *insertion loss*; i.e., how much it attenuates the desired frequencies. Filter quality is determined also by how greatly it attenuates the undesired signals, and over what range of frequencies. If the filter does not provide sufficient restriction of undesired energy over the stopband of interest, it is simply not adequate to the purpose, no matter what its other merits.

Having selected a filter to achieve the desired EMI control, there are yet other characteristic features which must be considered. For example, voltage and current ratings of the filter must be sufficient to allow operation under all expected circuit requirements. Large voltage deviations and steep transient pulses have to be accounted for, as well as all environmental conditions that the filter must withstand under prolonged usage. As part of the overall operating circuitry, adequate filter ratings and characteristics are essential to ensuring high systems reliability.

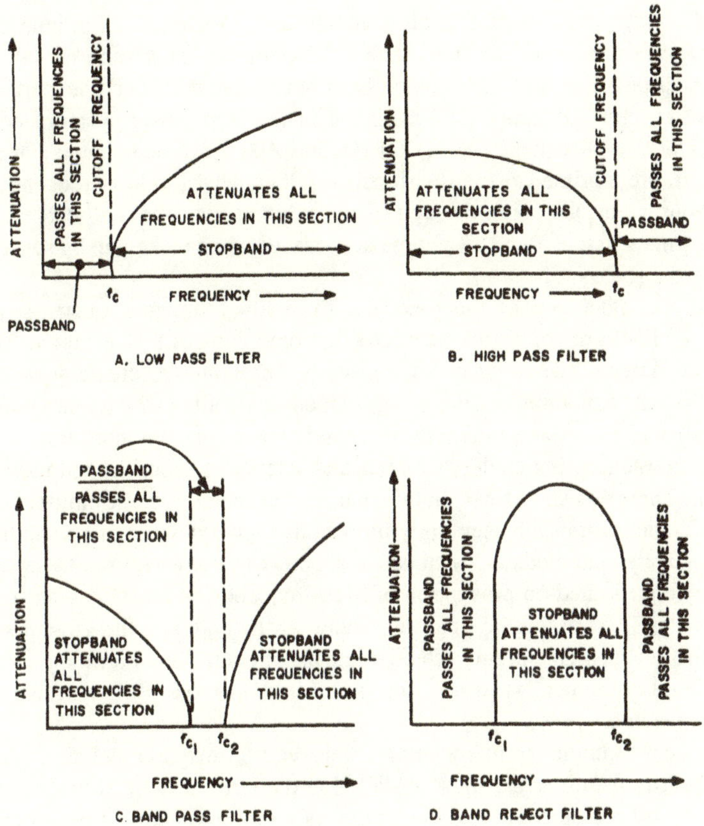

Figure 4-56 Functional Characteristics of Filters

4-2.4.2 Shipboard Filter Applications

A typical application of filters is in shipboard powerline circuits: 400-Hz main power distribution sources are commonly used aboard naval ships, along with interference-susceptible loads. Good filtering is a must to prevent pickup and conduction of harmonic interferences generated by power supply rectifiers. More significantly, many of the powerline EMI problems are associated with ground system currents flowing throughout myriad ship structures, including the hull, decks, framework, pipes, cable shields, conduits, equipment racks, and cabinets. Because of the variety of possible sources, these structure currents are complex and difficult to predict or measure. Nonlinear loads fed by the ship's primary power are frequently the source of ground currents. Variances in line-to-ground impedances create unbalanced line-to-ground voltages, hence differences of potential between ground points, to set up structure currents.

Harmonic product generation is another prevalent cause of ground current interference experienced in ship systems. Nonlinear loads such as solid-state rectifiers produce a high content of harmonics that then become part of the structure currents. Because they originate in the ship power system, structure currents are low-frequency (i.e., 60-Hz and 400-Hz fundamentals) and, left unfiltered, are a prime cause of performance degradation in low-frequency electronic equipment. Coursing through racks, cabinets, chasses, and cables, structure currents will make an unwelcome appearance when picked up by susceptible electronic circuitry.

At the other end of the spectrum from low-frequency electrical power sources of EMI are sophisticated microelectronic devices that generate RF disturbances. Digital switching in logic circuits, for example, create subtle interference signals containing harmonic components extending well up into hundreds of megahertz.[29] Coupled internally to chassis terminals, the interference easily reaches interconnecting cables, which act as antennas to conduct and radiate the harmonic energy as stray EMI. In this manner the cables become emitter sources of interference from such seemingly innocuous digital devices as personal computers, printers, and modems. Methods to suppress interference from these sources must be concentrated on prevention of high-frequency currents ranging from 30 MHz to 1 GHz from flowing into the circuit wiring and onto the ground shields of attached cables. One highly effective means of controlling harmonic radiation from digital equipment is to use low pass filters to block EMI currents at the cable connectors.

Recent examples of filters being employed routinely to preclude disruption of shipboard operations are those installed in the Central Control Station (CCS) lubrication oil pressure and level indicator monitoring systems of newer frigates and destroyers. In the frigates a single filter is inserted between the pressure indicator transducer output cable and the transducer body. In the tank level indicating transducer, two filter kits are used, one at the input and one at the output. In just these two types of CCS transducers aboard frigates there are 24 different filter configurations. Two samples are pictured in Figure 4-57.

The two primary sources of EMI which cause interference to the transducers are: (1) electrical broadband noise, and (2) HF communications transmissions in the 2–30 MHz band. Electrical broadband noise is created by the continual making and breaking of electrical contacts in nearby equipment and is characterized by high-intensity spurious products coupled onto the transducer cables. In the case of the HF transmissions, interference is coupled onto topside cables and conducted down to below-deck areas, where it is picked up by susceptible transducer cabling. In both cases the effect is to drive the transducer output signals into erroneous readings at the CCS, thereby causing false alarms.

INPUT SIDE (CONNECTED TO TRANSDUCER)

OUTPUT SIDE (CONNECTED TO CABLE)

(a)

INPUT SIDE (CONNECTED TO TRANSDUCER)

OUTPUT SIDE (CONNECTED TO CABLE)

(b)

Figure 4-57 Shipboard Pressure Transducer Monitoring System Filters

By using reactive components to impede interfering energy in series with the desired dc signals and to shunt the unwanted energy to ground, these filters prevent both the broadband electrical noise and HF interference from reaching the transducer circuits. An RLC double-pi network used in pressure transducer filters on destroyers is depicted in Figure 4-58.

Other everyday examples of filters used to avoid or control EMI in shipboard electromagnetic systems design are bandpass circuits of transmitter and receiver multicouplers; notch filters in EW equipment to suppress the fundamental frequencies of continuous wave radars (termed notch filters because they are extremely narrowband rejection filters used to exclude an unwanted fundamental frequency and to pass the rest of the band of interest); and band-pass–band-reject filters incorporated to protect IFF systems such as that illustrated in Figure 4-59.

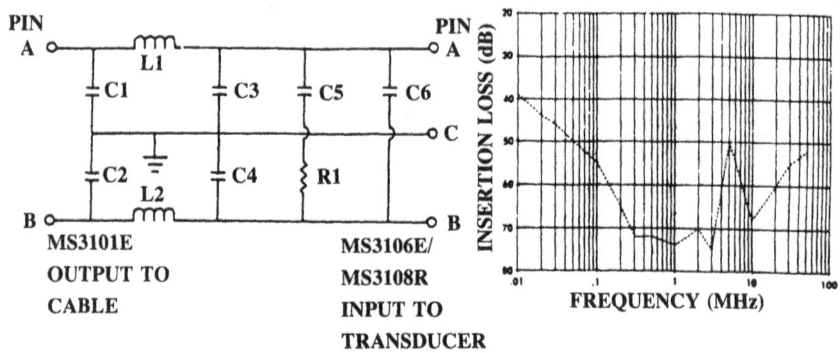

PART	DESCRIPTION
C1	0.02 MICROFARAD NPO DISC CAPACITOR
C2	0.01 MICROFARAD NPO DISC CAPACITOR
C3,C4,C5,C6	0.001 MICROFARAD NPO DISC CAPACITOR
L1,L2	6800 MICROHENRY INDUCTOR
R1	47 OHM, 1/2 WATT METAL FILM RESISTOR

NOTES:

1. Wire size 22 gauge stranded copper
2. Emerson & Cumming Potting Compound STYCAST #26S1, CATALYST #9
3. Tolerances should not exceed 10%

Figure 4-58 Pressure Transducer Schematic and Characteristics

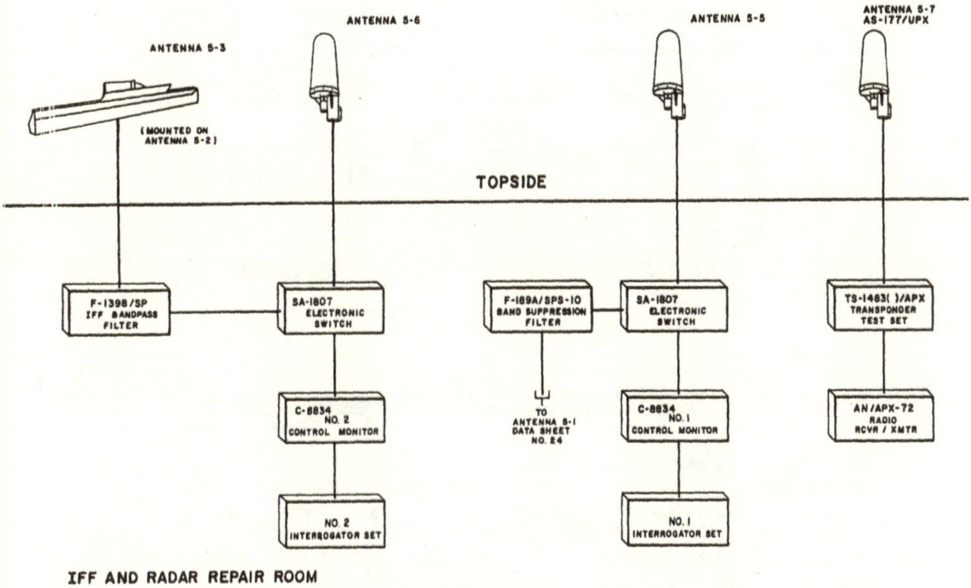

Figure 4-59 Filters Used to Protect Shipboard IFF System

4-2.4.3 Filter Installation Precautions

In addition to judicious selection of the correct filter to fulfill the required need, proper installation is essential to achieve the desired effect. In most cases it is best to place the filter in or on the apparatus that is generating the EMI; i.e., mount it at the source. It is important to establish as low as possible an RF impedance between the filter casing and ground. Consequently, the methods used to mount a filter become critical at high frequencies, where an improperly installed filter can result in impedances to ground sufficiently large to develop EMI voltages and to reduce the filter effectiveness. To maintain optimum bonding to the ground plane structure, both the surface on which the filter is to be mounted and the surface of the filter itself must be unpainted and thoroughly cleaned. Mounting ears and studs must ensure firm and positive contact to establish and maintain an RF impedance as close to zero as possible. Adequate separation of input and output wiring is imperative, particularly at high frequencies, as radiation from wires carrying interference signals can couple over directly to the output wires, circumventing the filter. Additionally, where chassis wall mounting isolation is not feasible, shielded wire should be used to assure adequate isolation. Figure 4-60 depicts various methods of correctly installing powerline filters.

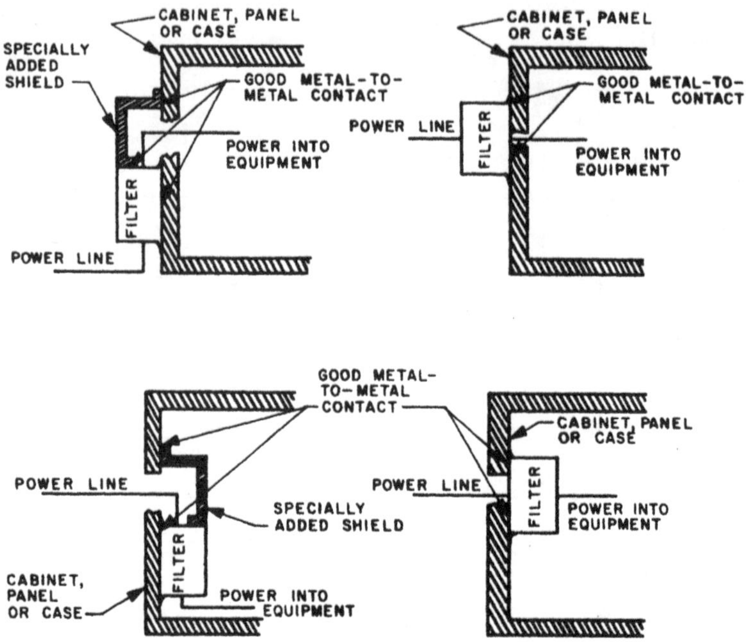

Figure 4-60 Powerline Filter Installation Methods

4-2.5 EMI Blanking Techniques

A unique method often used aboard naval ships to prevent reception of high-power local interference is a form of time domain synchronization called electronic blanking. This technique originated in the early 1950s when it was learned that ship EW passive intercept receivers were experiencing severe interference from onboard radar systems. Operating close by and simultaneously with the EW receivers, the radar transmitters emitted signals so intense that ordinary frequency domain practices such as filtering could not provide sufficient receiver protection, nor could the problem be eased by more careful selection of installation locations so as to provide adequate isolation between source and victim. In the limited topside volume available, the radars and EW intercept receivers simply could not be separated widely enough to preclude high-power mutual coupling of EMI.

It was evident that the only feasible alternative was to cut off, or "blank," the EW receiver at the moment of radar energy intercept. Since radar pulses are short in duration relative to the interval between pulse transmissions, there is adequate time to permit "look-through" of the EW receivers. By making use of energy supplied directly from the radar pulse circuitry to activate electronic gate switching, the Navy fabricated a blanking device to interrupt EW reception. The original 1953 experimental model, developed by the Naval Research Lab and named the AN/SLA-10 Blanker, is pictured in Figure 4-61. This five-stage filter using individual diode gates is the direct ancestor of highly sophisticated modern-day shipboard blanking equipment.

In later design philosophy it was found more expedient to have a pretrigger signal derived from each interfering source feed into a central blanking unit. The blanker generates and sends a blanking pulse to the victim receiver, interrupting its operation in synchronization with the anticipated interference emission. Present technology favors blanking at the intermediate frequency stages of the intercept receiver, circumventing the need for several frequency selective devices. In this manner all interfering signals are blanked, irrespective of individual frequency.

Figure 4-61 AN/SLA-10 Original Experimental Model (1953) Blanker (Photo courtesy of Naval Research Lab)

There is, of course, a distinct disadvantage of blanking: it interrupts, actually turns off, the receiver system operation for the duration of the interference. Carefully programmed synchronization minimizes the loss of reception (off-time) due to blanking intervals. Nevertheless, excessive blanking time can become a problem, particularly when a large number of onboard emitters cause various shipboard receive systems to be disrupted for a significant portion of their operation. As an actual example we will examine the case of a recent naval warship combat systems design (the case is representative, but by no means the severest). It will be seen that this blanking scheme involved complex electronic programming to ensure success.

Beginning with the preliminary design phase of the ship, efforts to achieve maximum EMC included establishing an EMCAB. One of the principal purposes of the EMCAB was to evaluate the total ship environmental effects. Within the constraints of topside volume (''real estate''), mission requirements, and means of controlling EMI, the ship topside design was continually refined and assessed.

Part of the EMCAB initiative included an analysis of each of the several emitter-receiver combinations in terms of direct and indirect EMI coupling. The analysis considered the ongoing topside design, documented fleet experiences, inspection board surveys, and EMI test and corrective action team reports. As a result of the analysis, solutions were identified and action was taken to preclude EMI between the many emitter and receiver pairs. For some of the directly coupled interference paths, blanking was determined to be the only viable solution, even though it was conceded that blanking is ultimately undesirable because it interrupts receiver performance.

In accordance with the EMCAB assessment, a precise blanking plan was developed. The plan required that blanking be provided for five onboard victim systems, including EW, air control radar, and telemetry data receivers. Eleven high-power radar and air navigation emitter sources, some with multiple operational modes and varying pulsewidths, would supply an aggregate of 15 pretrigger signals to the programmed input channels of a central blanker. The blanker system would determine which of the input pretriggers to combine into a series of blanking pulses for each of the five outputs, and would define the proper timing and duration for each output pulse. System off-time was also computed to summarize the effects of blanking for each of the victim receivers.

A final analysis of the blanking plan concluded that the effect of blanking on the performance of each of the five receive systems was within acceptable tolerances. It was noted, however, that in the case of one EW system, the blanking cutoff time approached one-fifth of the overall receive operational time. It was felt that any further blanking might cause that system to suffer noticeable degradation (a threshold compromise in that blanking prevents a more serious form of degradation). Blanking, the EMCAB acknowledged, is an EMI control method of last resort, to be applied only when no other solution is possible.

4-2.6 Topside Systems Arrangement Techniques

It should be apparent to the reader by this time that shipboard EMC involves many facets of systems engineering design, installation, maintenance, and EMI corrective practices. We have seen that achieving EMC requires an understanding of the shipboard EME with all its potential sources and victims of EMI. It requires a thoughtful management program and a vigorous EMI control plan. It requires thorough knowledge and proficient application of such EMI suppression techniques as shielding, grounding, bonding, filtering, and blanking either to prevent or to relieve system performance degradation. Notwithstanding all of the above, there is yet another crucial factor necessary to establish shipboard systems EMC. That factor, to be addressed now, is topside electromagnetic systems arrangement; i.e., the optimum placement in the ship topside of high-power emitters and ultrasensitive sensors to ensure both EMC and the effective reduction of EMI.

4-2.6.1 Antenna Interference Characteristics

Modern warships may have well over a hundred antennas. Simultaneously transmitting and receiving varied forms of information on frequencies as low as 10 kHz to above 30 GHz, each of the antennas is essential to the fulfillment of mission requirements. Because of the constraints in available topside space, however, it is difficult to select suitable locations so that the antennas may perform well.

The intrinsic electromagnetic nature of antennas is itself a major cause of the problem. An antenna functions at its best when well isolated from any of its own kind, from any other electromagnetic devices, and from any objects nearby which may interfere with good performance. Unfortunately, isolation aboard ship is virtually impossible, interference is everywhere evident, and antennas are peculiarly sensitive to interference in many forms.

Shipboard antenna interference may be categorized primarily as that due to blockage, coupling, RF emission, and high-level radiation.

a. *Blockage*—When an antenna must be placed near a large object such as a mast, or portions of the superstructure, or other antennas, a corresponding sector of its intended coverage will be shadowed out. If the antenna is to be used for receiving, that blocked sector is unusable and is either surrendered as such, or a second antenna is added to fill in the gap (complementary coverage). If instead the antenna is employed for transmitting, its radiated energy, unless prevented from doing so, will impinge upon the offending obstacle, causing reflections and scattering (and quite likely coupling and reradiation), thereby distorting, perhaps significantly, the radiation pattern.
b. *Coupling*—If the nearby obstacle possesses certain electromagnetic characteristics (for example if it is made of metal, or is another antenna), mutual coupling with this parasitic element will result, altering the impedance as well as the pattern. The effect on the system may be so great as to drastically reduce the antenna's utility.
c. *RF Emissions*—Reception of undesired emissions is one of the most frequently encountered forms of interference aboard naval vessels. It is the natural consequence of a relatively small platform crowded with so many radiators. Unwanted signals are generated on ship as harmonics, intermodulation products, noise spikes, and broadband noise, to be picked up by onboard sensors used for receiving distant, and often much weaker, signals.
d. *High-Level Radiation*—In many instances, particularly on warships, high-power radiators are required for carrying out ship missions. Often the emitters are microwave, posing biological hazards to personnel, but even at lower frequencies the high energy levels are a threat to onboard fuel and explosives. A common problem is that of HF transmitting antennas

inducing currents in nearby metallic structures to cause RF burns to personnel coming in contact. Restricted locations must be allocated aboard ship in which to place these high-power emitters so as to minimize the dangers of RF radiation.

Clearly, aboard naval ships, all the above conditions inherently must exist. There can be no true isolation of antennas. There will always be sources of interference close by which will adversely affect antenna (and, therefore, mission) performance. Consequently, it is this high potential for interference that the systems engineer must anticipate, and with which he must cope, during the design and integration process.

4-2.6.2 Preliminary Antenna Arrangement Considerations

The engineering problem facing the antenna systems designer is to place each antenna in the topside: (1) to provide good coverage, that is, to avoid blockage of the radiation pattern; (2) to realize maximum intended range for each antenna's purpose of communications, or navigation, or radar target search, detection, acquiring, tracking, illumination, and weapon control; (3) to avoid being susceptible to EMI; and (4) to avoid being a source of EMI or a radiation hazard to ship personnel, ordnance, and fuel. Only the most careful thought in placing the antenna can produce a topside integration which effectively achieves all of these objectives.

Shipboard antennas fall generally into one of three categories:

1. Omnidirectional antennas used mainly for communications, navigation, and passive reception to satisfy the need of ships and aircraft to maneuver independently of each other and fixed radio stations.
2. Directional antennas used for transmitting and receiving spatially concentrated energy in one direction at a time, e.g., radar, weapons control, and SATCOM, to radiate to or obtain information from remote sources.
3. Directional antennas used to determine bearing of incident radiation; e.g., direction finding, navigation, and EW.

To accommodate these three classes of antennas, four specific approaches are taken:

a. Broadband excitation of the masts and superstructure, as in the case of high-frequency, fan type, wire-rope antennas.
b. Probe excitation of ship structures as with VLF tuner whips; e.g., LORAN and OMEGA.
c. Tuned independent antennas such as 35-foot whips with base couplers.
d. Directional, independent antennas and arrays such as those used for radar and weapons control.

4-2.6.3 The Topside Systems Design Team

Designing the topside of a modern naval ship is an exercise of compromises involving several engineering competitors. Each competitor justifiably seeks to protect a special interest and to design the best possible subsystem within that sphere of interest. At the same time, however, each creates an impact, often severe, upon the other topside interests. Therefore, no competitor, whether weapons, navigation, communication, helicopter operations (helo ops), or other, should be allowed to optimize at the express disregard of others.

In the mid-1970s, it was recognized that a formalized procedure must be instituted to attain joint agreement among the several competing engineering elements represented in the design of a ship's topside. Accordingly, NAVSEA established the Topside Design Integration Engineering Team, or TDIET, to provide an engineering committee for new ship designs and for major modernizations and conversions. Regularly scheduled TDIET meetings bring together design specialists: principal engineers and architects from combat systems, hull structures, weapons arrangements, navigation, lighting, firing zone coverages, topside electromagnetics, mast design, stack exhaust dynamics, safety, and various other technical areas.

The TDIET is responsible for developing a topside systems arrangement that satisfies the ship performance requirements. The EMC systems integration engineer, as chairman of the TDIET, is charged with ensuring topside EMC, with EMI suppressed to the minimum effect possible. In this manner, the step-by-step derivation of candidate topside arrangements, including rationale for trade-offs and iterations, is documented, and the resultant design substantiated.

The challenge is to arrange each item in the ship's topside so that each will adequately meet its mission requirements, but in the early stages of design, the platform itself is generally undefined. Some relatively fixed boundaries, such as length and width of the proposed hull may be known, but not the height. Therefore, the topside volume is fluid and undefined. For example, it could not be known whether there will be one mast or two. Moreover, whether a mast will be a self-supporting pole or tripod or quadripod depends upon the number, size, and weight of the antennas it must support, some of which are massive. Other factors, such as bridge clearance restrictions, also affect mast height. The quantity of antennas to be mast-mounted may in turn depend upon what restrictions are placed on deck locations; i.e., deck zones must be kept clear for the firing of guns or the launching of missiles or the operating of helicopters, and for the replenishing of fuel or the handling of cargo. How high the antennas must be placed on the mast may be governed by such widely diverse factors as what radiation coverage must be provided, what weight and moment tolerances are allowed, and how much physical isolation from other antennas is required.

It is evident from this single example that no part of topside design can be done independently. Interaction with all other parts is imperative, and is, therefore, the *rasion d'être* for the TDIET.

The objective of the TDIET, as heart of the design process, is to reach agreeable compromises so that each topside element achieves satisfactory performance individually with minimum degradation to (and from) all other topside elements. The result must be a totally integrated topside system working in complete harmony.

Of course, the effort involves numerous trade-offs and iterations to arrive at alternative topside arrangement options. Detailed analyses are performed to predict weapon and electromagnetic performance, and at the same time to minimize EMI and radiation hazards. To aid in the analyses, computer modeling is done to determine systems performance factors such as coverage, blockage, and range. Radiation hazard restrictions are determined and an EMI maxtrix is developed to project potential sources and victims for the TDIET to mitigate (see Chapter 3, Figure 3-1).

Concurrently, to ensure and enforce EMC of the combat systems during the design process, the EMC systems engineer also chairs the EMCAB. Everyday debilitating or annoying EMI problems being experienced in the fleet, and the particular resolution applied, are fed back to the TDIET and EMCAB sessions. In this manner, the designers are able to avoid or correct deficiencies so that the EMI problems will not be perpetuated in the topside design.

4-2.6.4 HF Antenna System Integration

Given the proposed outlines of a new hull, the antenna designer is concerned immediately with the interrelationships of major topside items: the height and shape of the superstructure, placement of the deck weapon systems, location and form of the stacks, quantity and physical structure of the masts, and available installation space for antennas. At the early stages of topside design, none of the above are fixed. Placement of large, high-power HF antennas on deck will affect performance of the weapons, and vice versa. The quantity and weight of antennas proposed for mast mounting may determine the number of masts and will certainly influence the shape and height of any mast. Height and geometry of the superstructure above the main deck may influence greatly the radiation characteristics and impedance of certain antennas. Thus, each item affects the location and performance of the others. In the beginning, only gross arrangements can be suggested with alternatives proposed as options.

An obvious first step in the preliminary systems integration is a serious attempt to reduce the total number of antennas required. The most likely candidate is HF communications. Broadbanding and multicoupling are now used routinely to reduce the overall number of shipboard HF antennas. By dividing the HF band into three overlapping segments, namely 2-6, 4-12, and 10-30 MHz, and by using broadbanded antennas (e.g., wire-rope fans, trussed or twin whips), several transmitters or receivers may be operated into a single antenna. Not only is the number of antennas reduced but, since multicouplers are also filtering devices, electromagnetic interaction is lessened.

a. *HF Antenna Scale Modeling*—Using an antenna modeling range, scaled models of the ship with variable antenna configurations are subjected to carefully controlled measurements to determine the feasibility of the proposed arrangement or to recommend the best alternative. These models are usually $\frac{1}{48}$ scale, are made of sheet brass (see Figures 4-62 and 4-63), and include all topside structural elements influencing the HF antenna characteristics. Based on tests made on the range, changes to the model's topside may be made quickly and easily to expedite the HF antenna systems design.

The emphasis during modeling is on design of the broadband HF transmitting antennas, normally one each to cover the 2-6, 4-12, and 10-30 MHz frequency bands for each ship. The main objective is to provide, and integrate into the ship hull, HF communications antennas with efficient, omnidirectional radiation characteristics by attaining a 3:1 VSWR throughout the entire frequency band. Such efficiency is achieved by the physical form and resultant topside placement of the antenna itself, and by exacting calculations to derive the inductive-capacitive L-, T-, or pi-type matching network inserted in the transmission line at the antenna feedpoint for maximum transfer of energy. Through modeling experimentation, test, and analysis, the component values of the variable inductors and capacitors and their configuration in the matching network are so accurate that only final tuning adjustments need be done at the time of installation in the actual shipboard environment.

The antenna range, with its rotating lead sheet turntable simulating the sea "ground," is used also during modeling for radiation pattern measurements. Polar plots are made at specified band frequencies and varying elevation angle cuts over 360 degrees of azimuth. The resulting plots (see Figure 4-64) graphically illustrate the degree of coverage along with pattern nulls and perturbations of the proposed ideally omnidirectional antenna for a particular frequency and elevation radiation angle. Full-scale measurements at sea have over the years conclusively validated the scale modeling results.

The brass modeling design and testing of broadband HF receiving antennas is less demanding than for the transmitting antennas. Naval HF receiving antennas are not required to be highly efficient. On the contrary, it is

Figure 4-62 Scaled Brass Ship Model on Antenna Range

Figure 4-63 Scaled Brass Model Topside Antennas

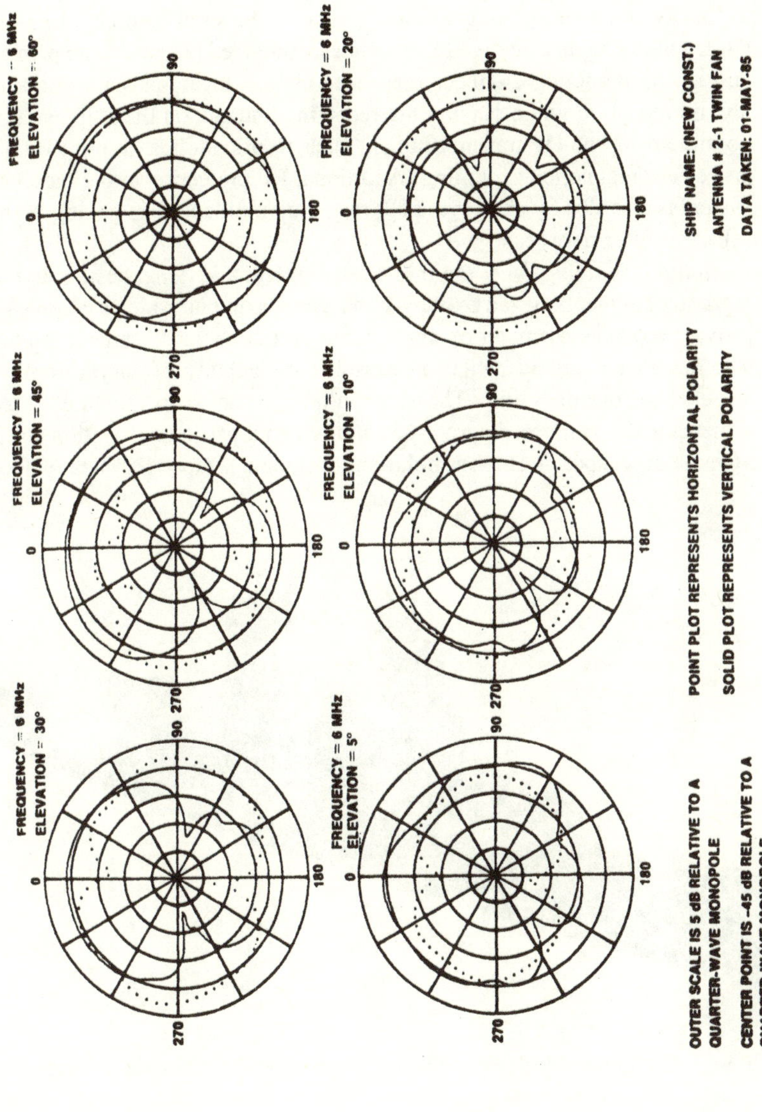

Figure 4-64 Modeled HF Broadband Antenna Radiation Pattern Plots

desirable that they be inefficient to the extent of just matching the threshold of the external minimum atmospheric noise level to the receiver system internal noise. The receiving antenna is thus made as insensitive as possible to locally generated high-level interference of the shipboard environment, while still being an effective receptor for its intended purpose. The primary aim of the modeling engineer then is to choose locations with good all-around reception while placing the receiving antenna on the ship as far as possible from the HF transmitting antennas to provide maximum isolation and electromagnetic decoupling. Additional RF protection to the receiving system is afforded by employing HF receiving multicouplers having highly selective filter networks.

b. *Computer Modeling*—By the mid-1980s, antenna systems design and integration engineers began making good use of computer-aided graphics to provide visual assessments of ship antenna and weapons placement rapidly. The designer is quickly able to determine the validity of varying options selected in topside siting. The ship hull is displayed in modular, three-dimensional form (see Figure 4-65) and, when a particular location for an antenna or weapon is chosen, polar and rectangular plots graphically show

Figure 4-65 Computer-Modeled 3-Dimensional Modular Ship Hull Form

the radiation coverage or firing zone coverage available. Likewise, blockage due to superstructure and deck obstacles is immediately evident from these computerized plots (see Figures 4-66 and 4-67). Using computer programs in this manner, the systems designer is readily able to illustrate those arrangements which are viable for further considerations. From there, additional computer-aided techniques are used to predict systems performance in terms of range, frequency, power, gain, and expected EMI to and from each emitter and sensor.

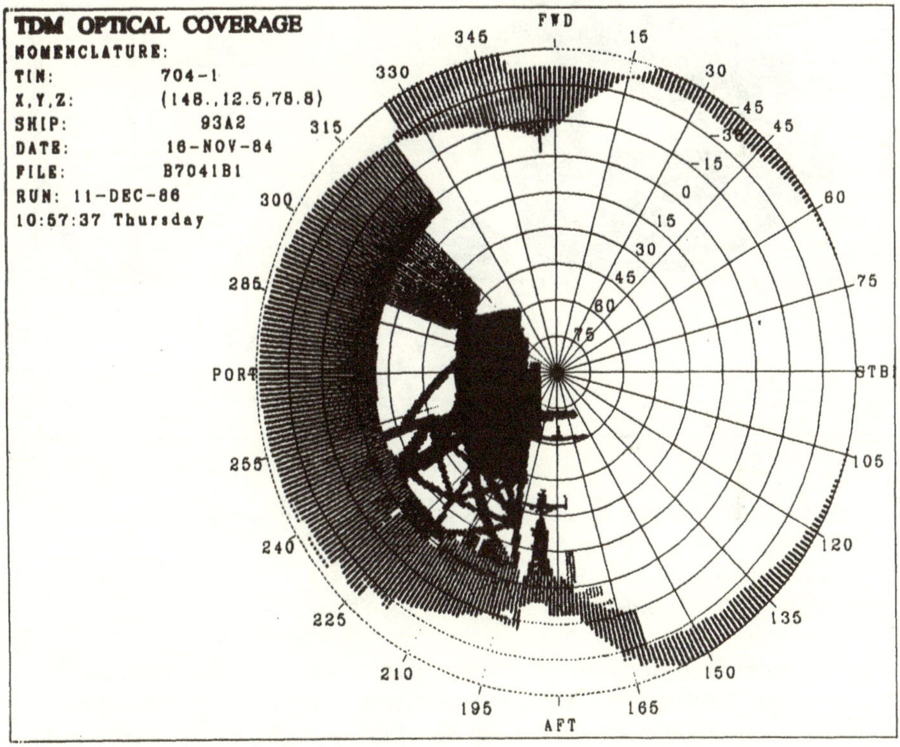

Figure 4-66 Computer-Modeled Sensor Coverage (Blockage) Polar Plot

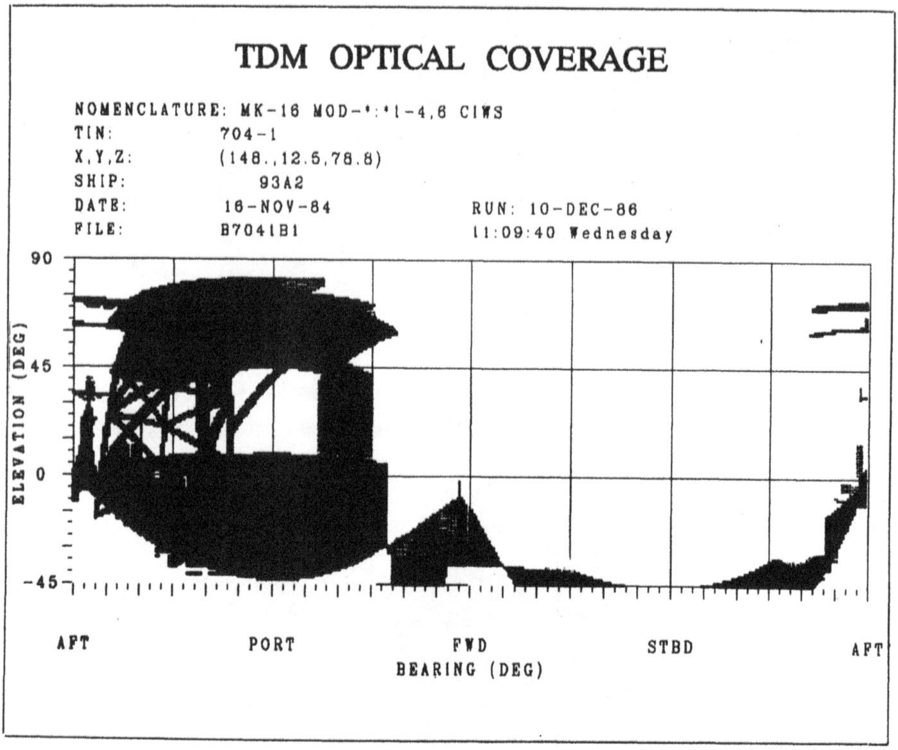

Figure 4-67 Computer-Modeled Sensor Coverage (Blockage) Rectangular Plot

4-2.6.5 EMC Considerations

Once a fairly firm complement of the number and types of antennas required is obtained, the next step in the topside design and integration process is to begin tentative placement of antennas and to anticipate the impact that the arrangement will have in terms of overall ship's predicted performance potential, EMI, and RADHAZ. In fact, it is this competition with other systems (and structures) that is most difficult to resolve. A first approach that might come to mind is to locate all antennas as high as possible, in the clear, for omnidirectional transmission and reception. The masts and yardarms would seem the best choice. Unfortunately, as seen in Figure 4-68, there are problems with this choice:

a. Communications engineers, weapons control engineers, radar engineers, navigation-aid engineers, and EW engineers all have the same hopes.

b. It is undesirable to collocate transmitting and receiving antennas in the same frequency band; one mode, either transmit or receive, has to be placed elsewhere. (Normally, the transmitting antennas are installed in the vicinity of the transmitter equipment room in order to minimize cable length attenuation losses.)
c. Some antennas, particularly transmitting antennas in the low portion of the HF band, do not function well when high above the water, their radiation patterns being apt to split up, forming multiple lobes in the elevation plane. To compound the problem even further, the yardarm and masts are used also to support flag halyards (which frequently become entangled in the antennas), commissioning pennants, navigation lights, and wind speed indicators.

As a result, only antennas that absolutely require such locations can be mast-mounted. For example, air-to-ground UHF communications antennas, TACAN, and direction-finding antennas are installed high above the sea to get the maximum possible line-of-sight range and to have an azimuthal radiation pattern which is as nearly circular as possible. For large, heavy antennas, other locations must be sought, and competition for real estate begins to get quite difficult: On any ship there are areas which are immediately eliminated; e.g., helicopter operation areas, vertical replenishment zones, gun arc-of-fire zones, missile launching zones, cargo and boat handling zones, and visual navigation zones. Additionally, antennas should not be installed on stacks or next to fuel handling areas or ordnance stowage areas.

Figure 4-68 Ship Mast Antenna Congestion

Isolation between antennas is maximized to the greatest extent possible. Separation of communications receiving antennas from high-power transmitting antennas is necessary to prevent overloading the receivers and the generation of intermodulation products within the receivers. Isolation not adequately afforded by physical separation is compensated by frequency separation, filtering, and blanking. It is also advisable, and in some cases essential, that isolation be provided between antennas of different functions; e.g., communications and radars, or search radar-to-navigation radar. A typical case is the requirement for UHF satellite communication antennas to be located well away from ship-to-ship UHF transmitting antennas.

4-2.6.6 Candidate Antenna Systems Arrangements

As a result of working closely with all the various engineering participants—hull, machinery, deck arrangements, weapons arrangements, and electrical—candidate topside configurations are proposed. The options fulfill each of the requirements to the greatest extent possible; it is, however, recognized that no single solution is capable of meeting all requirements. Trade-off studies determine those options most nearly meeting requirements, with the risks inherent in selection of each. Recommendations are made, with documented rationale for the selection, including the identification of any risks and deficiencies of the resultant system.

4-2.6.7 Post-Design Phase

During both the shipbuilding and the active fleet life of the ship, revisions are made ranging from simple additions of platforms and structural reinforcements to major changes in ship equipment complement. Such modifications more often than not will affect antenna characteristics, usually adversely.

Examples of topside changes which may seriously degrade antenna performance include addition of deck houses; extensions to bridge wings; modifications of mast and yardarm configurations; additions, deletions, or relocations of antennas; and changes in weapons systems. Since each antenna has been tailored to its specific environment, such alterations may have a dramatic effect upon topside EMC.

How well the total integrated shipboard electromagnetic system will function in the support of ship missions is only determined when put to the test of actual operations. At that time the quality of the overall topside design, including the initial planning, the model range studies, the EMC analyses, the coordinated iterative efforts to reach compromises, and the EM performance assessments, in derivation of the topside systems arrangement will be evident.

4-2.7 TEMPEST Electromagnetics

This chapter would not be complete without some consideration being given to a peculiar type of electromagnetic concern where, rather than noise interference, undesired emission of intelligence information must be entirely suppressed. The highly specialized discipline to accomplish this suppression is known by the short name TEMPEST. Not an acronym, TEMPEST is an unclassified term for the detection, evaluation, and control of conducted and radiated signals emanating from communication and data processing equipment. The concern is that such emanations might unwittingly reveal classified information being processed by the system equipment. The techniques used in TEMPEST practices to suppress electromagnetic emissions are generally equivalent to those used in the reduction of EMI; viz, shielding, grounding, bonding, filtering, and signal isolation. Therefore, the subject is of interest as an integral part of shipboard electromagnetics. Because of their classified nature, however, the details of TEMPEST test procedures, design techniques, and emission levels cannot be discussed here. Only a superficial overview can be presented for introductory purposes.

The Chief of Naval Operations document that implements national policy on the control of compromising emanations for facilities, systems, and equipment used to process classified information is OPNAV Instruction C5510.93. The Department of the Navy Supplement to DOD Directive S5200.17 addresses specific physical security criteria applicable to shipboard. MIL-STD-1680 translates these DOD instructions into installation criteria relevant to shipboard equipment and systems. Known as TEMPEST installation criteria, specific measures are required to minimize the possibility of compromising electromagnetic emanations.

The thrust of TEMPEST engineering is to restrict intelligence-bearing information to its proper identified zone—the so-called Red or Black areas in the system. Facilities, equipment, and circuits carrying and processing only unclassified plain language, or encrypted intelligence, are designated as Black. Components, circuits, equipment, and systems carrying uncrypted classified clear text are designated as Red and require strict protective control. Red zones are physically secured from unauthorized access. Obviously the encryption device itself is a Red/Black interface unit, and extraordinary precautions must be exercised to control access to Red intelligence information. These precautions include utmost control of conducted and radiated emissions. Unfortunately, a compromising path or source may be very subtle, requiring great effort to detect or suppress.

Isolation is the primary objective. Isolation between Red and Black information must be complete at every level in design, equipment, and system. The most notable sources of problems are power supply systems and the overall grounding scheme; so it is these that require most attention. Additionally, Red and Black cabling must be segregated and adequately shielded. Where feasible, fiber-optic cables are used to provide maximum isolation and energy decoupling. Other techniques applied to achieve sufficient isolation include: (1) filtering of Black lines that penetrate Red areas; (2) maintaining separation of Red and Black equipment racks; (3) separating control cables from power cables within equipment racks; (4) providing 360° bonding of cable shielding at the backshell; and (5) using double shielded outer braid on cables.

These are but a few of the generalized practices used in TEMPEST engineering. It can be seen that they exactly parallel those of shipboard EMI control. For specific technical details of TEMPEST, readers with an established need to know should consult the MIL-STD-1680 installation criteria document.

REFERENCES

1. "Electromagnetic Compatibility Program within the Department of the Navy," OPNAVINST 2410.31D, Office of the Chief of Naval Operations, Washington, DC, 6 August 1984, p. 1.
2. B. E. Keiser, *Principles of Electromagnetic Compatibility*, Artech House, Dedham, MA, 1983, p. 35.
3. *Electromagnetic Compatibility Installation and Maintenance Theory and Practice*, NAVSEA STD-407-5291779, Department of the Navy, Puget Sound Naval Shipyard, Bremerton, WA, March 1981, p. 2-2.
4. R. V. Carstensen, *Electromagnetic Interference Control in Boats and Ships*, Don White Consultants, Gainesville, VA, 1981, p. 1.3.
5. Keiser, *op. cit.*, p. 35.
6. "Electromagnetic Interference Reduction," *Electronics Installation and Maintenance Book*, NS 0967-LP-000-0150, Department of the Navy, Washington, DC, June 1972, p. 8-1.
7. G. C. Salisbury, "Analysis of Very High Order Frequency Components of the Intermodulation Interference Spectrum," Report TR 1852, Department of the Navy, Naval Electronics Laboratory Center, San Diego, CA, 26 October 1972, p. 5.
8. *The Electronics Material Officer's Guide to Shipboard Electromagnetic Interference Control*, NAVSEA STD-407-5287556, Department of the Navy, August 1980, p. 2-9.
9. Keiser, *op. cit.*, p. 113.
10. Carstensen, *op. cit.*, p. 5.25.
11. M. Smith and F. O'Conner, "New Shielding Potential with Metallic Fabric," *Interference Technology Engineers Master*, Robar Industries, West Conshohocken, PA, 1985, p. 334.
12. C. D. Storms, "EMI/RFI Shielding For Structural Foam Business Machine Components," *Interference Technology Engineers Master*, Robar Indus-

tries, West Conshohocken, PA, 1983, p. 53.
13. *Ibid.*, p. 54.
14. *Ibid.*, p. 54.
15. G. A. Krulik, "Electroless Plating For EMI Shielding," *Interference Technology Engineers Master*, Robar Industries, West Conshohocken, PA, 1984, pp. 180-188.
16. J. H. Ling, "EMI Shielding: Selection of Materials for Conductive Coatings," *Interference Technology Engineers Master*, Robar Industries, West Conshohocken, PA, 1986, p. 186.
17. W. Gindrup and R. R. Vinson, "A Logical Approach To EMI Shielding Materials," *Interference Technology Engineers Master*, Robar Industries, West Conshohocken, PA, 1986, p. 184.
18. *Electromagnetic Compatibility Design Guide for Avionics and Related Ground Support Equipment*, NAVAIR AD1115, Department of the Navy, Washington, DC, July 1980, p. 5-30.
19. G. Olenskyj, "Selection of EMI Shielding Gasket Materials," *Interference Technology Engineers Master*, Robar Industries, West Conshohocken, PA, 1984, p. 62.
20. E. J. Carlson and W. S. V. Tzeng, "Galvanic Corrosion Measurements On Various EMI-Gasketed Joints," *Interference Technology Engineers Master*, Robar Industries, West Conshohocken, PA, 1985, p. 312.
21. Olenskyj, *op. cit.*, p. 62.
22. *Handbook of Shipboard Electromagnetic Shielding Practices*, NAVSEA S9407-AB-HBK-010, Department of the Navy, Washington, DC, 1982, pp. 6-1 through 6-26.
23. *Ibid.*, p. 2-1.
24. *Hull Generated Intermodulation Interference Reduction Techniques for Forces Afloat*, NSO967-266-1010, Department of the Navy, Washington, DC, October 1971, p. 31.
25. *Shipboard Bonding, Grounding, and Other Techniques for Electromagnetic Compatibility and Safety*, MIL-STD-1310D, Department of the Navy, 8 February 1979.
26. R. S. Key, "Kevlar Lifelines," *Deckplate*, Vol. 6, No. 4, July-Aug 1986, p. 17.
27. D. F. Herrick and M. S. Geoghegan, "The Effects of Life Rails on Radar Antenna Performance," Final Technical Report Task B029, Naval Surface Weapons Center, Dahlgren, Virginia (Circa 1985).
28. A. W. Laine, Ph.D., "EMI Filters," *Interference Technology Engineers Master*, Robar Industries, West Conshohocken, PA, 1985, p. 244.
29. H. Hazzard and R. Kiefer, "EMI Reduction Using Capacitive Filter Connectors," *Interference Technology Engineers Master*, Robar Industries, West Conshohocken, PA, 1986, p. 240.

Chapter 5

Shipboard Electromagnetic Radiation Hazards (EMR)

5-0 THE RADIATION HAZARDS PROBLEM IN GENERAL

In recent years there has been a rapidly increasing public awareness of potential biological harm resulting from unwitting exposure to electromagnetic radiation. The problem has been especially dramatized in the case of microwave frequencies. Although public anxiety at times has been fostered by alarming reports in the media, the desire for clear and forthright information on the subject of RADHAZ nevertheless is justified. We live in an electronic age, immersed in an electronic environment, and, while reaping the material benefits, we must be kept well informed of possible adverse consequences.

Affluent societies such as ours make use of virtually the entire electromagnetic spectrum from powerline energy frequencies to x-ray frequencies. The dominant high power usages for the civil population are electrical utilities and radio and television broadcasting. Numerous other heavy demands on the spectrum include: (1) commercial airlines air-to-ground communication, navigation, and air traffic control radar; (2) law enforcement communications and radar surveillance; (3) commercial shipping, recreational flying, and boating communications and navigation radar; (4) medical specialized equipment for detection and treatment of disease; and, perhaps of greatest significance, (5) military radar and communication networks that dot and web the landscape from shore to shore. Even in the home such radiating devices as microwave ovens and automatic remotely controlled garage-door openers are now commonplace. As a result, the general public throughout the industrialized world is quietly (and invisibly) being subjected to continuous, increasing levels of electromagnetic radiation (EMR). The question is, do these radiation exposure levels pose a threat to health? More importantly to our subject matter, how do we preclude exposure to EMR from being harmful to shipboard personnel?

Our special interest here is in the biological effects of nonionizing RF radiation—that is, radiation other than that from, say, nuclear fusion reactions, weapons containing nuclear warheads, or from intentional medical use of such ionized energy as ultraviolet and x ray. In distinguishing between ionizing and nonionizing radiation it should be pointed out that, in the portion of the electromagnetic spectrum above the infrared segment, energy from ultraviolet rays, x rays, and gamma rays reacts with living matter with such force as to ionize, or electrically charge, organic molecules. In so doing, chemical bonds are destructed, causing damage to tissue and possible disruption of biological functions.[1] Nonionizing radiation, on the other hand, at the microwave and lower frequencies lacks sufficient intensity and concentration of energy to ionize organic material, but still may interact in an elusive, indirect manner. The effects of radiation for our purpose therefore will be understood to mean the results of nonionizing radiation. Of particular concern is the problem of electromagnetic energy of such intensity as to affect (1) human tissue, (2) flammable fuel vapors, and (3) explosives; i.e., to cause biological damage, ignition of fuel, and detonation of ordnance aboard ships.

5-1 BIOLOGICAL EFFECTS OF RADIATION

Electromagnetic energy impinging upon a human body may be reflected, absorbed, or transmitted through the tissue, or some combination of these.[2] The biological result of this contact is the subject today of extensive study, debate, and controversy. Remaining unsettled are precisely which frequencies, what energy levels, what radiation conditions, and what mechanisms actually cause interaction. And, given that an interaction occurs, which biological effects are harmful, which are perhaps even beneficial, and which are harmless or ineffectual?[3] In fact, a biological effect very likely may have no significant health consequences. Only when it produces injurious or degrading alterations to the health of an organism is the effect a biological hazard. The degree of harm depends upon such interrelated factors as the frequency of radiation, energy intensity, polarization of the field, and duration of exposure.

Intensive investigations over the years have established that biological damage to living tissue will result from penetrative heating if critical levels of radiated power density and length of exposure are exceeded. Such thermal damage affects vulnerable body parts, including the skin, muscles, brain, and central nervous system; the effect is most severe, however, for delicate organs with little ability to dissipate heat, such as the lungs, liver, testes, and portions of the eye. Furthermore, radiation levels may be so low as to cause no apparent harm to tissue, yet be adequate to raise the whole-body temperature or to generate localized hot spots within the body.[4] In such incidences physiological control mechanisms of critical body functions may suffer.

The body's ability to dissipate heat successfully is dependent upon such factors as the ambient temperature and air circulation rate of the environment, clothing being worn, power density of the radiation field, amount of radiation absorbed, and duration of exposure to radiation. Temperature regulation in the body is accomplished primarily through sweat gland evaporative cooling and by heat exchange in peripheral circulation of the blood. The regulation process is complex and adverse effects produced when high temperatures are induced in the body may result in decreased system efficiency. Because of the body's limited ability to lower heat through perspiration and blood circulation, only a moderate increase above normal temperature can be tolerated. Where areas of the body are cooled by an adequate blood flow through the vascular system, there is less likelihood of tissue damage resulting from abnormal temperature; in body areas having relatively little blood circulation, however, the temperature may rise considerably from lack of means for heat exchange. Consequently, biological effects of EMR are more likely to occur where there are radical rises of temperature.

Under moderate conditions physiological changes seem to be tolerated by the body's normal capability to adjust and correct. The fear is that when the body is unable to make compensating adjustments to radiation overheating, lasting harm is done. In other words, thermal damage to tissue may be irreversible in those cases where the body is unable to replace the tissue through natural process, resulting in lasting side effects.

The human eye is a case in point. Certain parts of the eye's vascular system are inefficient for the circulation of blood and the exchange of heat to the surrounding tissues. The lens of the eye in particular appears to be very susceptible to thermal damage. Unlike other cells of the body, the transparent lens cells of the eye cannot be renewed. When the cells making up the lens become damaged or die, opacities or cataracts develop. The loss of transparency is usually a slow process and the individual begins to suffer impaired vision. It can be readily appreciated, therefore, why there is such concern for preventing radiation overexposure to sensitive organs of the human body.

More disturbing even than thermal effects are recent revelations that various nonthermal problems may be observed from experimental microwave radiation tests; i.e., in certain circumstances chromosomal damage in live animals is being reported.[5] Some experts are concerned that these and other biological changes break down the body's immune systems, cause behavioral changes, and promote the development of cancer. Moreover, there is growing evidence that 60-Hz powerline electrical and magnetic fields produce biological effects in humans, albeit as yet apparently not harmful.[6]

Is there a danger, then, of being exposed to EMR, and, if so, how much and where? To date, the results of thousands of studies over the past 40 years seemingly confirm that despite intense use of the entire electromagnetic spectrum, the general public is in no danger of being exposed to harmful levels of radiation. The US Government's Environmental Protection Agency (EPA) has performed investigations at nearly 500 locations in 15 cities that show, even in those environments subjected to high-power radio broadcast and television transmissions, that 95 percent of the population is exposed to extremely weak levels of radiated power density; that is, no higher than 0.1 microwatts per square centimeter.[7] (Weak fields are described as those radiation levels that do not produce temperature increases in animals above normal body fluctuations: in general, power densities below 1 milliwatt per square centimeter (mW/cm^2) over a frequency range of 30-300 MHz. Moderate levels of 1-5 mW/cm^2 are tolerated by human beings for short duration.[8])

The EPA studies correlate well with many others, all of which indicate that the American public, although living in an environment filled with myriad forms of EMR, is exposed to energy levels hundreds of times below current US guidelines of safe, permissible intensity levels. Based on these data it might be concluded that nonionizing radiation poses no threat of harm to the general public in today's highly industrialized society. However, as has been pointed out earlier, the debate and controversy go on and will perpetuate until the findings are no longer inconsistent, inconclusive, or ambiguous, even though "no clear cut damage to human beings from low-level radiation has been demonstrated."[9]

There are, though, two classes of people known to be subject to potentially hazardous levels of EMR: occupational workers and military personnel. In the case of occupational workers there is a large variety of apparatus in use that radiate electromagnetic energy. These devices include microwave food processors; industrial plastics heat sealers; chemical analysis equipment; medical diathermy, detection, and therapeutic equipment; science and research laboratory equipment; radio and television broadcast equipment; and microwave telecommunication transmitting equipment—all of which should require adherence to federal, state, and local regulations to avoid potentially harmful levels of radiation exposure. These concerns, however, are outside the realm of our particular interest. We wish to address the specific problem of protecting military personnel from the hazards of EMR.

5-2 SHIPBOARD HAZARDS OF ELECTROMAGNETIC RADIATION TO PERSONNEL (HERP)

We have seen in previous chapters that, with an extraordinary density of high-power emissions, the shipboard environment is conducive to the generation of EMI. It is likewise true that the large number of high-power emitters radiating highly concentrated energy in and around so confined a platform makes naval surface ships among the most potentially hazardous of electromagnetic environments in which people must live and work.

Quite aware of the severity of the problem and, desirous to prevent any chance of overexposure to its shipboard personnel, the Navy began in the 1950s to establish and enforce safe radiation exposure limits. Little was really known at the time of the nature and effects of EMR interacting with the human body. But, anxious about exposure to high-power microwave fields in particular, each of the military services was eager to support research and experiments which would aid in the derivation of guidelines to ensure adequate protection of personnel. Industry, science, and other Government agencies also felt the need of setting standards.

5-2.1 Origin of Radiation Exposure Limits

As an outcome of studies done in 1953 at the University of Pennsylvania, the first tentative recommendations for safe radiation exposure limits were made. Projecting the anticipated results of heating organic tissue and of possible damage from overexposure, and incorporating a safety factor of 10, a power density of 10 mW/cm^2 was proposed.[10] The US Navy quickly accepted this limit. It was applied to all frequencies between 100 MHz and 100 GHz without any restriction on duration of exposure. In 1966, the American National Standards Institute (ANSI), a private organization which publishes voluntary regulations, formally issued the 10 mW/cm^2 limitation in its ANSI C95.1 Standard. When, in the mid-1970s, ANSI refined the standard to constrain the exposure duration to 6-minute intervals and lowered the frequency of interest to 10 MHz, the Navy complied immediately. This new standard gained wide acceptance in the United States as the single most important nonionizing EMR exposure standard. It remained the Navy's accepted HERP level for 30 years, until in the 1980s it was abruptly revised.

5-2.2 Emergence of Modern Radiation Exposure Standards

During the three decades from the 1950s to the 1980s there was relatively quiet acceptance of and adherence to the ANSI power density radiation exposure limits. However, as we noted, among the public there began to arise sharp interest, concern, and controversy regarding just what constitutes safe levels of exposure. Continuing studies only added fuel to the debate. Publicity served to bring the issue to the attention of worried politicians and high-level officials throughout the Government. The interest and visibility did not go unnoticed by either the scientific community or the military.

Reflecting the mood of concern for public safety, ANSI, in its 1982 periodic review of standards, updated the C95.1 guidelines to account for current theory of energy absorption. Improvements in radiation exposure measurement techniques over the years had resulted in a determination that the energy absorbed by animal tissue is a direct function of radiation frequency, polarization of energy (i.e., orientation of the electromagnetic field components), and physical size of the irradiated body. For example, a normal size adult human standing in a vertically polarized radiation field is an efficient receptor of electromagnetic energy, due to body resonance, in the 70 MHz to 100 MHz VHF range—a low range that was not included in the original power density exposure limits adopted by the Navy during the 1950s and early 1960s.

ANSI, therefore, realizing the *frequency-dependence* of exposure intensity, radically modified its standards to limit the *absorption* of radiation energy. Careful to keep its former safety factor of 10, ANSI offered a new C95.1-1982 standard of 0.4 W/kg of whole body weight averaged over any 6-minute period, and, at the same time, imposed stricter exposure limits (lower power density levels) above 10 MHz. Accordingly, for the body resonance frequencies of 30 to 300 MHz, the limit is now much more restrictive, 1 mW/cm^2, a value one-tenth the previous level. Also, the overall frequency range of the standard was expanded by extending the lower frequency end to 300 kHz. To differentiate this new concept from the old, the exposure limit terminology was changed from the former "power density" to the current "specific absorption rate," or SAR, in the derivation of frequency-dependent permissible exposure levels (PELs) of radiation dosage. Note how this concept of time rate of absorption per mass of tissue now correctly concedes that the potential for EMR hazards includes frequencies well below microwave.

5-2.3 Shipboard Permissible Exposure Criteria

The Navy, embracing the 1982 ANSI philosophy of frequency-dependent rate of energy absorption, proceeded to make a few modifications to the new standard so as to establish its own preferred safeguards. The radiation frequency range was broadened still further to cover the electromagnetic spectrum all the way from 10 kHz to 300 GHz; i.e., from VLF to infrared. Additionally, two separate categories of exposure level allowances were defined: (1) *restricted access*, which, based on criteria developed by the American Conference of Governmental Industrial Hygenists (ACGIH) as an occupational standard, applies to ships at sea and excludes persons less than 55 inches in height, and (2) *unrestricted access*, which conforms to the ANSI standard and applies to the general public irrespective of body size, and includes ships in port. The exposure limits for these two categories are identical up to 100 MHz, but differ widely above 100 MHz.

In adopting the new SAR exposure limits, and in keeping with its policy to avoid unnecessary risk of EMR to personnel (or, when risk is unavoidable, to ensure that any exposure is within safe limits and as low as reasonably achievable), the Navy issued its new radiation exposure criteria on 30 July 1985 as OPNAV Notice 5100.[11] Directed as a uniform, Navy-wide protection criterion, the Notice requires application to all phases of equipment design, acquisition, installation, operation, and maintenance. Tables 5-1 and 5-2 itemize the various PELs as averaged over any six-minute interval for both the restricted and the unrestricted access categories, respectively. Figure 5-1 depicts the exposure levels in graphic form.

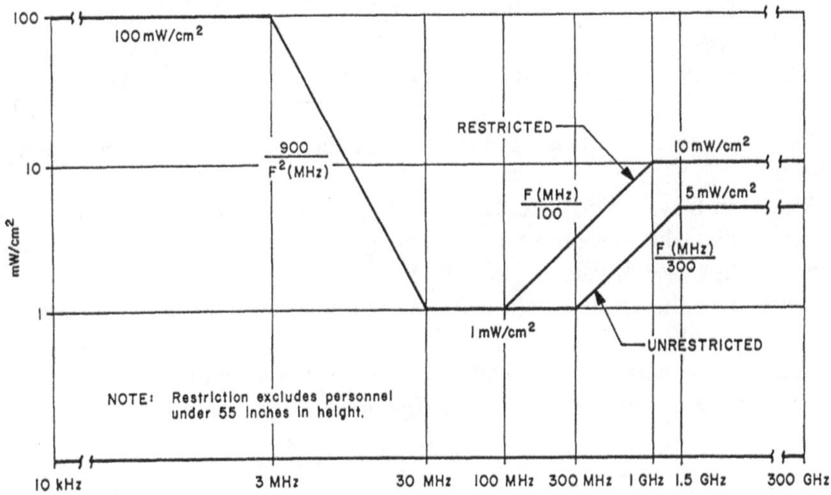

Figure 5-1 Whole Body Radiation Hazards Exposure Limits

There are some special exceptions to the normal PELs, as stated below:

a. Personnel who, as patients, undergo diagnostic or therapeutic procedures in medical or dental treatment facilities are excluded.
b. Devices operating at or below 1 GHz with an output power of 7W or less are excluded.
c. The derived PEL criteria in Tables 5-1 and 5-2 may be exceeded under special circumstances, provided it can be demonstrated by measurement that:

    1. The whole-body SAR does not exceed 0.4 W/kg when averaged over any 6-minute period.
    2. The spatial peak SAR (hot spot) does not exceed 8.0 W/kg averaged over any one gram of body tissue.
    3. The peak electric field intensity does not exceed 100 kV/m.
    4. Personnel are adequately protected from electric shock and RF burns

Table 5-1. Equivalent Permissible Exposure Levels for Restricted Areas[1,2,3,4]

Frequency (MHz)	*Power Density* (mW/ cm^2)	*Electric Field Strength Squared* (V^2/m^2)	*Magnetic Field Strength Squared* (A^2/m^2)
0.01–3	100	400,000	2.5
3–30	$900/f^2$	4,000 $(900/f^2)$	0.025 $(900/f^2)$
30–100	1.0	4,000	.025
100–1,000	$f/100$	400 $(f/100)$	.025 $(f/100)$
1,000–300,000	10	40,000	.25

[1]Restricted access areas are controlled to exclude persons less than 55 inches in height.
[2]Values in these tables were derived using the impedance of free space of 400 ohms. This value is rounded up from the generally accepted value of 377 ohms to allow for ease of calculations. Also, f is in MHz.
[3]When both the electric field and magnetic field are measured, use the more restrictive value.
[4]Tables apply only to whole body exposures and are based on the overall SAR of 0.4 W/kg averaged over 0.1 hour (six minutes).

through the use of electrical safety matting, safety clothing, or other isolation techniques.

The power density PELs listed in Tables 5-1 and 5-2 are derived for far-field plane wave conditions and apply only where a strict far-field relationship between the electric and magnetic field components exists. In the Fresnel zone near fields (such as for HF communications transmitting antennas aboard ship) both the electric and magnetic field strength limits, rather than the power density values, must be used to determine compliance with the PELs. Furthermore, it is important to note that in all cases, exposure levels must never exceed an electric field maximum intensity of 100 kV/m.

As part of the shipboard safety measures, radiation hazard warning signs are required at all access points to areas in which exposure levels may be exceeded. The format of the signs follows that suggested by the ANSI C95.1-1982 standard as shown in Figure 5-2. The warning symbols consist of black wavefronts radiating from a stylized point source antenna on a white background enclosed in a yellow and red hash-bordered triangle. (This color scheme deviates somewhat from the ANSI standard in order to ensure proper awareness of the

Table 5-2. Equivalent Permissible Exposure Levels for Unrestricted Areas[1,2,3,4]

Frequency (MHz)	*Power Density* (mW/cm^2)	*Electric Field Strength Squared* (V^2/m^2)	*Magnetic Field Strength Squared* (A^2/m^2)
0.01–3	100	400,000	2.5
3–30	$900/f^2$	4,000 $(900/f^2)$	0.025 $(900/f^2)$
30–300	1.0	4,000	.025
300–1,500	$f/300$	4,000 $(f/300)$	.025 $(f/300)$
1,500–300,000	5.0	20,000	.125

[1]Unrestricted access areas are not controlled and all persons may enter.
[2]Values in these tables were derived using the impedance of free space of 400 ohms. This value is rounded up from the generally accepted value of 377 ohms to allow for ease of calculations. Also, f is in MHz.
[3]When both the electric field and magnetic field are measured, use the more restrictive value.
[4]Tables apply only to whole body exposures and are based on the overall SAR of 0.4 W/kg averaged over 0.1 hour (six minutes).

sign in all shipboard lighting conditions, from low-level red light to yellow sodium vapor light.) For areas where access to radiation levels greater than 10 times the PEL may exist, warning signs are to be considered insufficient to ensure adequate protection. Instead, additional warning devices and controls such as flashing lights, audible signals, and various physical constraints such as guardrails and interlocks are required to prevent the chance of overexposure.

5-2.4 Shipboard EMR Hazards Protection Techniques

The requirements of OPNAVNOTE 5100 to protect naval personnel against overexposure to nonionizing radiation from 10 kHz to 300 GHz must be implemented in every US Navy surface ship and in all new ship designs. The immediate shipboard problem, in order to abide by the more stringent radiation exposure limits, is to ensure maximum safety to personnel with minimum adverse effect on ship mission operations within the existing topside systems arrangement. This is first met by enforcement of the restricted access criteria: ships at sea, after all, are operated under strictest access control and do not have personnel less

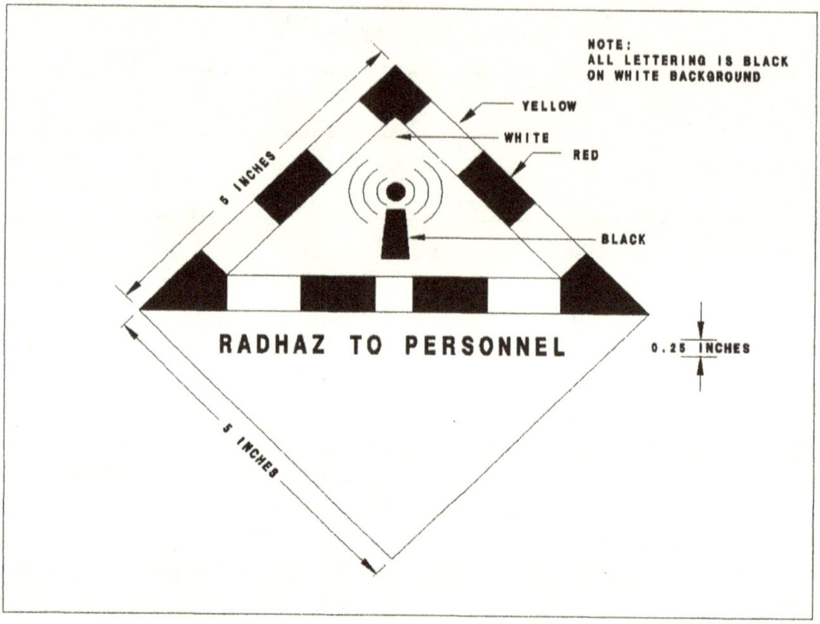

Figure 5-2 ANSI Standard Radiation Hazard Warning Sign

than 55 inches tall aboard. Nevertheless, in the performance of routine operational duties crew members do work in and around areas where high-power transmitting antennas are installed. Some of these antennas, particularly those used for HF communications, radiate high-level energy omnidirectionally throughout the whole topside. Therefore, such commonly mannned open deck areas as lookout and watch stations, replenishment-at-sea stations, signal and search light positions, catwalks, and passageways are within the field of radiation. Isolation by spatial separation is simply not a viable solution in so limited a platform volume. Finding satisfactory solutions is a continuing challenge.

Important first steps are to make sure that all shipboard personnel are well informed about potential radiation hazards, and to enforce exposure limits. Warning signs must be posted, and danger zones must be clearly marked by circles painted around all transmitting antennas that pose a threat to safety. Radiation hazard advisories must be issued which specify safe operating conditions for allowable levels of exposure to personnel working in the area.

The next step involves a detailed electromagnetic mapping of surface ship topsides through specialized radiation level environmental surveys.[12] In this manner source emitters and areas of excessive exposure level are identified as potentially hazardous. Unsuspected sources of hazardous radiation, such as energy leakage from waveguide flanges or faults and from malfunctioning radiation cutout devices which allow radiation past the limits, may be detected as part of the survey. Procedures for controlling personnel exposure are then determined.

The final step requires implementation of specific methods to reduce the potential for creation of hazardous conditions. These techniques include the relocation of antennas, reduction of transmitter power, careful management of operational frequencies, and erection of nonmetallic liferail barriers around hazard zones. Systems engineering techniques to reduce the likelihood of radiation hazards to personnel must be applied as discussed below.

5-2.4.1. Ship Design Criteria to Control EMR Hazards

The foremost naval ship design element that determines the EME is the placement of and relative proximity among the several transmitting antennas. During the process of attaining the optimum arrangement of the antennas, the radiation exposure concerns are considered in balance with all the other leading design needs such as weapons arrangement, combat systems performance objectives, operational and mission effectiveness, and weight and moment constraints. Designers keep in mind all the while that the permissible exposure levels of Table 5-1 must be adhered to during operations at sea, and those of Table 5-2 when in port or when carrying passengers less than 55 inches in height. In fact, the design goal of all new naval ship construction programs will be the more stringent exposure levels of Table 5-2.

As part of the topside systems design various sources of information are used. Shipbuilding specifications, ship drawings, equipment technical manuals, technical reports, ship operations personnel, and onboard topside surveys all will assist in determining the following design factors:

- Types of electromagnetic systems to be installed.
- Maximum on-axis power density and PEL distances from each emitter.
- System operational requirements of all potential EMR hazard sources.
- Location of all potential EMR hazard sources and the relationship to all normally occupied areas.
- Types of radiation-limiting mechanisms, present or proposed settings, and methods of override.
- Ship design characteristics for pitch and roll.

For the purposes of EMR analyses, shipboard emitters are classified as stationary, rotating, or directed beam. The design requirements differ somewhat for each category.

a. *Stationary Emitter Design Requirements*—Stationary emitters generally are used aboard naval ships for HF, VHF, and UHF omnidirectional communications transmissison. HF antennas are the most difficult to cope with in systems EMR design as they must be installed in areas that preclude physical contact by shipboard personnel. VHF and UHF antennas ordinarily are placed high in the superstructure or on the ship masts to provide clear all-around transmission. Since these systems generally are line-of-sight and therefore are of low gain and radiating low power, they are not normally an EMR problem.
HF transmit antennas (radiating in the military 2 MHz to 30 MHz band) must have a minimum 4-foot horizontal and 8-foot vertical physical clearance in all occupiable topside areas. HF whip antennas that radiate more than 250 watts must provide a 12-foot minimum physical clearance radius from any portion of the antenna to any occupiable topside area.
b. *Rotating Emitter Design Requirements*—Shipboard rotating antennas include several types of 2-D and 3-D air search, surface search, navigation, and air-traffic control radars, some of which radiate extremely high levels of pulsed energy. These antennas are usually mounted on a platform high on the ship mast or on a pedestal in nonoccupiable areas. Because they rotate they produce intermittent exposure and are therefore seldom an EMR problem. Nevertheless, the topside design must confirm that no radiating antenna will be placed in a location that will cause the PEL to be exceeded in any normally occupiable area, particularly when it might be in its highest duty cycle mode or allowed to radiate in a nonrotating mode.
c. *Directed Beam Emitter Design Requirements*—Weapons control radars, SATCOM transmit antennas, and EW emitters are examples of directed beam antennas used in shipboard systems. Weapons control radar antennas are very high gain and have very narrow radiation beamwidths for concentration of high energy. They are pitch- and roll-stabilized, and because of weight and performance considerations are frequently mounted lower than rotating emitters and are therefore near to normally occupiable areas. Most of these radar antennas therefore employ radiation cutout devices such as mechanically operated switches, computer software, or mechanical stops which prevent radiation into selected areas. SATCOM and EW transmit antennas have similar characteristics and require like design considerations. Directed beam emitters are the most common source of EMR problems, and their problems can be the most difficult to resolve.

For design purposes, directed beam antennas must be placed as high above normally occupiable areas as is electrically and mechanically practical still to satisfy all other system performance requirements. An EMR cam cutout scheme must be developed for each directed beam antenna to prevent the irradiation of any normally occupiable area with EMR levels that exceed the PEL. At the same time, directed beam antenna coverages must not be degraded by an EMR cutout device that does not have an override (battleshort) capability.

5-2.4.2 EMR Hazards Measurements and Analysis

EMR field measurements are required to be taken aboard naval surface ships regularly to ensure the safety of shipboard personnel. The purpose is to determine that permissible levels of radiation exposure are not exceeded in normally occupiable areas near ship radar and HF transmit antennas, beyond the limits of radiation cutouts, or from RF leakage in below-deck compartments. In all events the risk of overexposure must be prevented. Overexposure for Navy purposes is defined as any human exposure to nonionizing EMR that exceeds the permissible exposure level by a factor of five. Overexposure requires immediate medical attention, and a report of the incident must be submitted within 48 hours.

The most commonly used test instruments for taking RF field measurements are those that have broadband isotropic monitor probes integral with a radiation level indication meter.[13] The isotropic probe allows near-equal response to energy arriving from any direction except along the instrument handle. Hence it is not necessary to rotate the probe in any manner to strive for a maximum reading. Power from the electromagnetic field under test is dissipated in the isotropically spaced thermoelectric elements of the probe's lossy media. A low-level dc voltage is subsequently generated and conducted to the instrument preamplifier. Typical EMR meters require two probes to cover a frequency range from 10 MHz to 26 GHz, and measure average power flux densities varying from 0.2 mW/cm^2 to 200 mW/cm^2. Meter response time, the time needed for the meter to reach 90 percent of its steady-state reading, can introduce a significant error if the radiating antenna under test is rotating or scanning, or if the test probe is moved quickly through a narrow radiation beam. If either movement is too rapid the meter will not have time to reach its full value and will indicate too low a reading. A meter response time of less than 1 second is preferred, and must not exceed 1.5 seconds.

a. *Preliminary Test Procedures*—Prior to commencement of actual test measurements, a thorough inspection of the test areas should be conducted. Waveguide systems above deck and in radar equipment compartments should be checked for loose flange bolts, cracks, or other faults that might allow escape of energy. The condition of flexible waveguide sections requires special scrutiny. Where energy leakage is suspected, the locations should be noted for follow-up tests.

   For directed-beam emitters that use radiating cutout devices to prevent illumination of selected safe zones, the cutout mechanisms should be checked so as to be sure that they are functioning in accordance with the set limits. HF transmit antenna installation sites should be examined to confirm that they incorporate required EMR safety practices. Likewise, all areas subject to EMR overexposure should be inspected to make sure that hazard warning signs are posted properly. Finally, shipboard administrative safety procedures for the operation and checkout of potentially hazardous radiating systems should be examined to certify that they conform to specified requirements.

Personnel performing the EMR tests must avoid any possibility of radiation overexposure. If the measurements are to be taken while the ship is dockside, there must be sufficient clearance in the vicinity of the ship to preclude EMR exposure to persons on the dock and on adjacent ships and piers. Test engineers should at all times be thoroughly familiar with the anticipated power density levels at the various field points to be tested. Navy technical manual OP 3565 lists the PEL distances and maximum allowable exposure times for various operating modes of shipboard emitters.[14] The distances and times are based on exposure to mainbeam radiation, though it is seldom necessary for test personnel to be in the full power mainbeam while taking EMR measurements. To be sure that excessively high levels of radiation are not present, EMR hazard monitors should be used to quickly check the test area. When excessive levels are detected, the transmitter power should be selectively reduced and the test data then extrapolated in the same ratio.

In almost all cases shipboard EMR tests are performed under simulated operating conditions, with the emitter deliberately stopped at or pointed to desired azimuth and elevation angles. The test location and length of time for personnel to take the measurements while the antenna is radiating should be predetermined. Judicious choices of these test conditions are necessary as they must be representative of those which could be encountered during wide variations of actual operations. Measurements made under unrealistic test conditions could result in critical EMR overexposure situations going undetected, unreported, and uncorrected. Therefore, even maximum pitch and roll conditions should be simulated so as to test to the worst case operations.

Test point locations should be selected which typify normally occupiable areas in the topside, such as on bridge wings, at flag bags, at signal searchlights, and at lookout stations—wherever it is reasonable to assume shipboard personnel would be present.

b. *EMR Test Measurement Guidelines*

1. *Stationary Emitter Tests*—The primary objective of this portion of the shipboard EMR survey is to determine the maximum power density in HF transmitting areas and the PEL contours in normally occupiable

spaces nearby. This is to be done for each HF transmitting antenna over its range of operating frequencies. The following general process is recommended:

(a) Energize the appropriate transmitter to full power using a mode of operation that produces maximum average power output. Record the power output reading. (If full power is not achievable, use reduced power and extrapolate test values by the same ratio.)

(b) In the selected test location, take measurements at heights of 6 feet, 4.5 feet, and 3 feet; i.e., at the approximate head, chest, and genital heights of an upright human body.

(c) Search for and record the highest E-field and H-field levels throughout the test area. If any measured values exceed the allowable PEL for continuous exposure conditions, repeat the tests to determine whether the criteria for intermittent exposure are also exceeded.

Note: Large metallic objects near HF transmit antennas will capture and reradiate electromagnetic energy. It is possible in some cases that the reradiated energy level will exceed the PEL. It is then necessary to measure and record the maximum power density and distance of occurrence around the reradiating object to ascertain at what point the reradiated energy level drops below the PEL.

2. *Rotating Emitter Tests*—Since rotating beam radiators move continuously, they normally do not present a risk of overexposure. Therefore, by authority of Navy manual OP 3565 [14], radiation tests need not be conducted for rotating emitters unless specifically directed by official request. When required, the following general process is recommended:

   (a) Energize the appropriate transmitter to full power (or specific reduced power as recorded) using an operating mode that produces the highest duty cycle.

   (b) Take measurement data at selected locations, stopping the antenna rotation at the point of maximum power density level.

   (c) Search for and record the highest power density levels in the test area. If any test value exceeds the allowable level for continuous exposure, repeat the tests to determine whether the criteria for intermittent exposure is also exceeded.

3. *Directed-Beam Emitters*—Antennas in this category are capable of producing highly concentrated on-axis maximum, average power densities from 200 to 400 mW/cm^2. These power densities may cause excessive exposure levels at any location subject to illumination by the radiation beam. In fact, EMR overexposure levels can extend to distances several

hundred feet beyond the ship. Special concern must be exercised, therefore, for the safety of individuals on nearby piers and ships during the tests.

Most weapons control radar antennas of the directed beam category have some form of radiation cutout circuit or mechanical stops that are set to prevent radiation into specified areas. Some of these antennas have just two azimuth settings and one elevation setting to avoid radiation into a selected azimuth sector and into areas above or below a desired elevation. This arrangement, it is important to note, seldom permits the radar to operate under maximum roll conditions without an EMR overexposure occurring at some point. Some weapons radars which have computer-controlled radiation cutout circuits can be programmed to allow numerous settings in both azimuth and elevation. With this capability it is generally possible to obtain contoured cutout zones around the ship to provide optimum EMR protection while affording acceptable system performance. It is recommended that, as part of the EMR testing, measurements be taken to verify that original radiation cutout settings are still correct, or, where necessary, to reset to new limits as a compromise between system performance effectiveness and personnel safety. The following general process is recommended for directed-beam radiation level testing:

(a) Determine the relationship between each weapons control radar (or other directed-beam antenna) and normally occupiable areas so as to choose test measurement locations properly; epecially note all potential overexposure areas.
(b) Provide for test positioning of the antenna either manually with azimuth and elevation hand cranks, or electrically from an operator's console. Make sure that all personnel are safely clear as many such antennas are capable of rapid acceleration and may inflict serious injury. Measurements to determine the power density in selected locations are to include situations where the antenna is depressed to low angles equivalent to those which result from actual pitch and roll of the ship. It may be necessary in such tests to bypass cutout limit switches and temporarily disconnect the ship gyro information input.
(c) Train the antenna to the test location bearing, using an elevation which maintains mainbeam radiation several feet above the test location and test engineers.
(d) Prepare for initial EMR measurements by holding the test probe at a height of approximately six feet above the deck.
(e) Energize the radar system and slowly lower the elevation angle until the PEL reaches the allowable level (per Figure 5-1) for the

frequency under test. Move the probe from side to side and up and down to verify that the reading is maximum for that particular test site. Record the azimuth and elevation angles and the test data. If unexpected high energy levels are found to be the result of reflections from various shipboard objects, document those situations which might prove hazardous under normal operating conditions.

4. *Waveguide Emission Leak Tests*—Checks for the escape of microwave energy from waveguides should be made to determine the locations of leakage, the radiation levels, and whether possible overexposure to personnel may have previously occurred. In the event of possible overexposure, it is important to document whether the PEL has been exceeded in occupiable areas, and whether personnel were in the area long enough to have suffered overexposure. Measurements should be carefully performed to determine at what body height the waveguide radiation leaks occur.

c. *EMR Test Measurement Analysis*—Power density tests conducted on ship must take into account a number of variables so as to minimize the chance of measurement errors. Test locations, for example, are frequently at a point where the complex electromagnetic fields under measurement are extremely irregular. In many instances the tests are made within a few feet of an emitter, and are therefore in the radiation near-field (Fresnel region) so that the measured energy fluctuates widely from point to point. Part of the radiation energy may arrive directly from the antenna mainlobe or sidelobe, or it may combine with energy reflected from the deck, the bulkheads, the masts, the stacks, or even the body of the test engineer. If these energy components were to all combine additively the test meter would indicate inordinately high levels. If, however, the direct and reflected energy components were to interfere with one another destructively, the meter readings would be quite low. In reality the combining of energy fluctuates somewhere between these two extremes, and, for that reason, averaging techniques must be applied, as we will see.

The two classifications of EMR exposure on ship are: *continuous* and *intermittent*.

1. *Continuous Exposure*—A continuous exposure EMR environment is one in which an individual may experience a constant level of radiation exposure for six minutes or more. For this case the frequency-dependent PEL is given in Figure 5-1. Typical shipboard radiation sources that produce continuous exposure levels are HF transmit antennas and waveguide energy leaks. Personnel required to stand watches or operate systems at fixed locations for periods in excess of six minutes must be

made aware of the potential for reaching the continuous exposure criteria, although because of a person's movement about the area it is unlikely that actual continuous exposure conditions exist aboard ship.

Nevertheless it must be stressed that, in a continuous exposure environment, levels in excess of those shown in Figure 5-1 are not acceptable. Emission sources producing excessive radiation levels must be reported and documented, technical or operational procedures must be initiated immediately to prevent overexposure of personnel, and the area must be clearly identified as an EMR overexposure danger zone.

2. *Intermittent Exposure*—An EMR situation in which an individual may be exposed to varying levels of radiation during a six-minute interval is known as an intermittent exposure environment. Most EMR exposure environments (with the exception of those caused by HF transmitting antennas and waveguide leakages) are intermittent. Generally it is quite difficult to determine with certainty the average exposure received by a person during a six-minute period because of movement around the area by the individual, movement (rotation or scanning) of the antenna radiation beam, and variations in power output levels. One recommended procedure for estimating the exposure for a person in an intermittent radiation situation is the time-weighted-mean average method. To use this method power density and time measurements are taken at each of several locations where a person is exposed during a six-minute period, so that

$$E_W = \frac{P_1 t_1 + P_2 t_2 + \ldots + P_n t_n}{t_1 + t_2 + \ldots + t_n}$$

where

E_W = Time-weighted average exposure for six-minute intervals in mW/cm^2

P = Equivalent plane wave power density measured at each specific location in mW/cm^2

t = Time at each specific location, and $t_1 + t_2 \ldots + t_n \leqslant 6$ minutes

Several measurements should be taken at each of the specified locations to determine EMR exposure levels at head, chest, and genital heights (i.e., approximately 6 feet, 4.5 feet, and 3 feet). The average of the readings recorded at each height is then used in the E_W equation for each of the locations and time periods. If E_W exceeds the PEL for the area, immediate technical and operational procedures must be initiated to prevent personnel overexposure, and the area must be clearly identified as an EMR overexposure danger zone.

Because the radiation beam of shipboard air search, surface search, aircraft control, and navigation radars is constantly rotating or scanning, any resulting EMR exposure is intermittent. Rotating radar antennas which have relatively wide beams allow longer intervals of exposure per revolution but produce less concentrated energy than that of narrowbeam radars. The average power density from a rotating radar antenna is approximated by

$$P_{AR} = \frac{B_W P_F}{360}$$

where

P_{AR} = Average equivalent plane wave power density at a point within the axis of rotating radar mainbeam in mW/cm^2
P_F = Fixed power density in mW/cm^2
B_W = Radiation pattern beamwidth in degrees

(It is interesting to note that rotation rate does not enter into the above equation. At a given test point if the rotation rate of an antenna is changed by a factor, Δ, the exposure time to the main beam at that point for one rotation is changed by $1/\Delta$. When the rotation rate is less than the averaging time base for a PEL of six minutes it is not a function in determining the P_{AR} of rotating radars.)

Similarly, for scanning beam radars the average power density is approximated by

$$P_{AS} = \frac{2B_W P_F}{A_S}$$

where

P_{AS} = Average equivalent plane wave power density at a point in the scan beam in mW/cm^2

P_F = Fixed power density in mW/cm^2
B_W = Radiation pattern beamwidth in degrees
A_S = Scan angle in degrees

Navy manual OP 3565 [14] gives the PEL distance for fixed beam, rotating beam, and scanning beam antennas for naval shipboard radars. With the exception of some low-power navigation radars, rotating and scanning beam radar antennas are usually placed on mast platforms well above normally occupiable topside areas. However, in the event that the mainbeam of rotating and scanning emitters might illuminate a normally occupiable area while locked on target or otherwise stopped, test measurements should be taken to document any potential for overexposure. If the exposure levels are found to exceed the PEL, technical or operational procedures must be implemented immediately to preclude overexposure to personnel, and the area must be clearly identified as an EMR overexposure danger zone.

One very important case of intermittent exposure possibility in a normally occupiable area is that of directed-beam radiation during ship pitch and roll. A good example is a weapons control radar director electrically trained in azimuth and elevation to track or illuminate a target automatically. Recall that because of their great size and weight, weapons radar antennas ordinarily are installed close to and only slightly higher than normally occupiable areas. Ship gyro information is fed to the antenna servo system to stabilize the antenna against the effects of pitch and roll. Generally, weapons radar directors cannot track a target or be pointed very much below the horizon line. When the radar beam is on the horizon, however, added pitch and roll can actually reduce the antenna-to-deck angle sufficiently to allow mainbeam irradiation of deck areas. Consequently, since pitch and roll cause intermittent conditions of exposure, and weapons radar directors use scanning beams, the exposure level for these antennas is computed from the

$$P_A = \frac{2B_W P_F}{A_S}$$

average power density equation given above for scanning beam radars.

d. *EMR RF Burn Hazards*—RF burn is a unique shipboard personnel hazard caused by EMR in a congested multisystem environment. Distinct from radiated power density exposure or electrical shock, RF burn is a natural consequence of the coupling of nearby HF transmit energy into topside metal items such as stanchions, king posts, liferails, crane hooks, booms, rigging, pipes, and cables. Upon casual contact with these RF-excited metal objects, an individual experiences an involuntary reaction to the alarming burn or spark. The contact voltage is itself neither lethal nor severely dangerous, but the uncontrolled response may well result in serious bodily injuries from reflex actions of falling away or striking other objects when in close quarters.

Regardless of their intended use, all metallic items have electrical properties of resistance and inductance. A third electrical feature, capacitance, exists between the items. The magnitude of these electrical properties depends upon the nature of the metallic material, the size, shape, and physical orientation of the objects, the proximity of the objects to each other, and the degree of grounding and bonding to the main ship structure. The effects of the inductance and capacitance vary with frequency and can be roughly simulated by the simplified equivalent circuit of Figure 5-3, which shows a relationship of shipboard king posts and cargo booms.[15]

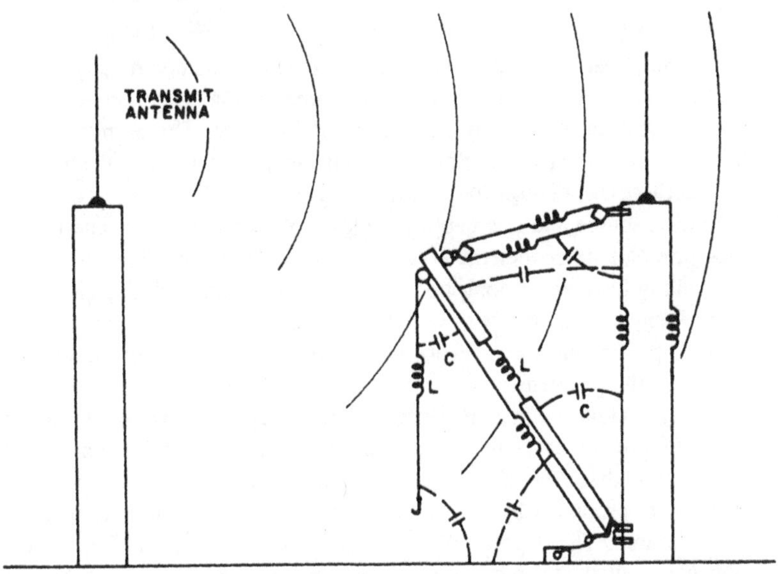

Figure 5-3 Electrical Equivalent of Shipboard Cargo Handling Equipment

The electrical circuit characteristics inherent in metal structures can act to intercept electromagnetic energy so that currents and voltages will be developed in the circuit impedances. At high frequencies the reactive components are significant, and when the inductive and capacitive reactances are equal (at resonance), maximum voltages will occur. The behavior is similar to that of a communications receiving antenna; in fact, metal objects which have the physical and electrical characteristics of HF receiving antennas are quite commonplace in shipboard topsides. Long metallic items are very efficient interceptors of RF energy. Some typical shipboard deck items and their resonant HF frequencies are:

- Antisubmarine Rocket Launchers — 12 MHz
- 3″/50 Gun Barrel — 14 MHz
- Underway Replenishment Stanchions — 4 MHz
- 35-Foot Metal Poles — 6 MHz
- A-4 and F-4 Aircraft — 6, 9, or 18 MHz (depending on the orientation of the aircraft with respect to the antenna)

Shipboard HF communications antennas radiate vertically polarized fields; therefore, vertical stanchions, pipes, king posts, masts, booms, davits, and cables readily couple the RF energy. The amplitude of the coupled energy is a function of (1) the length of the metal object with respect to the radiation frequency (wavelength), (2) distance between the radiating source and interceptor, (3) level of radiated power, and (4) physical orientation between the polarized field and the interceptor. Cargo ships especially have many long booms, king posts, and cables, and thus have high incidences of RF burn hazards. However, cargo ships are not alone; any ship carrying high-power HF transmit antennas is very likely to experience the problem.

An actual RF burn is caused by an RF current flowing into a person coming into contact with (or near enough to create a spark from capacitive coupling with) an electromagnetically excited metal object. The burn occurs from heat produced by the flow of current through skin resistance in the contact area. The degree of heat ranges from warm to painful. However, the exact level at which contact with an induced RF voltage should be classed as an RF burn hazard is not absolute. Experience has shown, for example, that severe burns can occur with the small contact area of a single finger, whereas with the entire hand at the same point of contact the effect may be unnoticed.

For Navy purposes, hazardous RF burn levels are defined as those voltages which cause pain, visible injury to the skin, or involuntary reaction. The term "hazardous" does not include voltages so low as to cause only annoyance, a stinging sensation, or moderate heating of the skin. The Navy has resolved that an open circuit RF voltage on an object in an EMR field in excess of 140 volts is to be considered hazardous. The 140-volt level is based on tests and measurements which indicate that a person will receive an RF burn when coming into contact with that voltage.

Misunderstanding of the causes of RF burn is evident from reports of the problem. The most common misconception is that the voltage builds up like static electricity, and is caused by improper transmitter operation. In fact, voltage appears instantaneously when energy is intercepted from the transmitting antenna. It remains only as long as the energy is being transmitted. The amplitude of the induced voltage is proportional to the square root of the radiated power; thus, a properly tuned transmitter will induce higher voltages than a poorly tuned one.

As part of naval EMC engineering practices, RF burn should be eliminated from ships as completely as possible. Several techniques are currently available, including the following:

1. *Hook Insulators*—Fiberglass filament-wound insulator links installed between metal cables and cargo hooks are very effective in preventing RF burns when contact is made with the hook itself. (The RF voltage and potential for burn remains hazardous above the insulator links, of course.) Examples of an uninsulated cargo hook and an insulating link used in a cargo hook are shown in Figures 5-4 and 5-5, and lightweight insulators in deck tiedown hardware are shown in Figure 5-6. Heavy-duty insulator links such as those in Figure 5-4 are available with capacity ratings of 15 tons, 30 tons, and 50 tons.
2. *Nonmetallic Materials*—Use of nonconductive materials for fabrication of such items as lifelines, guardrails, stanchions, jackstaffs, and posts has proven very effective for elimination of RF burn voltages. Recall, too, from Chapter 4 that use of nonmetallic materials greatly reduces the generation of intermodulation interference.
3. *Antenna Relocation*—One of the principal causes of RF burn formerly occurring in cargo ships was the common practice of installing HF transmit whip-type antennas high atop metallic king posts. The RF energy generated along the king post and coupled into nearby booms and rigging was intense. It is now standard practice never to install HF transmit antennas high on stanchions or superstructure items where an individual can come into contact.

   Similarly, there are times when HF antennas have had to be relocated in order to eliminate RF burn hazards. In most cases a minimum separation of at least 50 feet is required to achieve adequate reduction of coupled HF energy below the hazardous level.

Figure 5-4 Uninsulated Cargo Hook

Figure 5-5 Insulated Cargo Hook

Figure 5-6 Insulator Links in Tiedown Hardware Linkage

4. *Operational Procedures*—In some instances an RF burn hazard can be eradicated only through the use of operational restrictions. These include operating transmitters at reduced power output levels, use of alternative operating frequencies, avoiding simultaneous use of transmitting antennas, and avoiding HF transmission during various deck activities such as cargo handling or replenishment at sea. Effective operational procedures usually can be formulated only after careful tests and analyses.
5. *Burn Gun Measurements*—After considerable investigation and experimentation over many years, the Navy successfully developed a test instrument commonly referred to as a burn gun to detect potentially hazardous RF burn voltages. Integral to the instrument is a meter that indicates the RF voltage between a metallic object under test and the hand of the individual holding the gun. Ideally the voltage level registered is a good indication of whether a person would receive a burn if the object under test were touched. In reality, however, whether the measured voltage will cause a burn is largely dependent upon the impedance of the circuit being tested. The impedance can be compared to an internal power source impedance that determines the ability of the circuit to sustain the voltage and to deliver sufficient power to produce a burn. The burn gun has proven to be a reasonably good indicator of burn

Figure 5-7 Use of Burn Gun to Detect RF Voltages

probability in RF hazard surveys conducted aboard ship. It has been of inestimable value in initiating corrective steps to alleviate RF burn situations. Figure 5-7 shows test engineers using the burn gun to measure RF voltages on the metal surfaces of a ship weapons director.

5-3 HAZARDS OF ELECTROMAGNETIC RADIATION TO FUEL (HERF)

Perhaps nothing strikes fear in the hearts of seafarers like the report of a fire onboard. News reports many times over have recorded ghastly scenes of runaway fire damage on the decks of aircraft carriers. As recently as the 1982 Falklands Campaign the public viewed the charred remains of the once sleek British warship *HMS Sheffield* fighting desperately to stay afloat. Because of the concern for fire, ship crews are frequently and systematically drilled in the practice of firefighting and safety procedures. Again, as a result of the peculiar nature of multimission operations in a crowded ship, another EMR hazard is present, known as HERF. A large part of Navy shipboard practices to avoid the causes of fire is in awareness and preclusion of HERF.

The possibility of having fuel vapors ignite accidentally by metal-to-metal arcing created from high EMR fields aboard ship has been the subject of extensive study and research.[16] The probability for accidents is highest during fuel handling operations that take place near high-power transmitting antennas. Laboratory experiments and shipboard tests have shown that, while it is possible to ignite volatile fuel-vapor mixtures by induced RF energy, the probability of occurrence during fueling procedures is remote. Several conditions would have to exist simultaneously in order for combustion to be initiated:

- A flammable fuel-air mixture must be present within range of the induced RF arcing.
- The arcing must contain a sufficient amount of energy to spark ignition.
- The gap across which the arc would occur must be on the order of a half-millimeter.

Knowing that these conditions must exist has led to HERF control practices to reduce the likelihood of accidental ignition:

- Care in topside systems design to install HF transmitting antennas in sites well away from fueling stations and fuel vents.
- Use of pressurized fueling systems incorporating additives to preclude the formation of fuel-air mixtures at 1 atmosphere on aircraft aboard ship.
- Use of JP-5 fuel in almost all cases for aircraft aboard ship.

Still, even though the potential for HERF has been reduced by the above practices, it is yet present when handling the more volatile fuels aboard ship such as JP-4, aviation gas (AVGAS), and motor vehicle gasoline (MOGAS). When handling these fuels, personnel must be made fully aware of EMR hazards and the importance of following safety precautions.

5-3.1 The Nature of HERF Combustion

Under normal operating conditions the handling of gasoline does not produce a flammable atmosphere except close to vents, at open fuel inlets, or close to spilled gas. When air moves, as with wind across the deck in nearly all cases of ships under way, the fuel vapor is diluted and rapidly dispersed, greatly reducing the possibility of ignition. The flammability of fuels is also influenced by the fuel temperature. If the temperature is too high the hydrocarbon vapor content is likewise too high (i.e., too rich a mixture) for good ignition. If the temperature is too low, the hydrocarbon vapor content is too little (i.e., too lean a mixture) to support good combustion. Therefore, each fuel has a characteristic range of temperature, that is, a flammable hazard range, where the vapor-air mixture is best suited for combustion. Approximate typical high combustion temperature ranges for some of the fuels used aboard ship are:

- AVGAS: −40° to +10°F

- JP-4: −40° to +70°F
- Kerosene: +110° to +165°F
- JP-5: +130° to +210°F

Naval laboratory tests have concluded that arc energy is a determining factor for ignition of fuel vapors, and a threshold value of 50 volt-amperes is needed to cause gasoline to ignite. Using this criterion, measurements have been made for various fueling scenarios, relating the 50 volt-amperes to electric field intensity, radiated power, and distance from the EMR source antenna. From this work a general guidance graph has been derived using a typical HF conical monopole transmitting antenna, as shown in Figure 5-8 (from [14]).

Lab experiments also have determined that a minimum spark gap of about 0.02 inch (one-half millimeter) is required for ignition of a fuel-air mixture. In the case of shipboard fueling operations, metal-to-metal contact would have to be abruptly separated (making and breaking of contact) to create tiny half-millimeter spark gaps in a high intensity EMR field to draw a spark of sufficient length and energy to ignite fuel vapors. Consequently it is extremely important to ensure that static ground wires, tiedown cables, and other metallic connections to aircraft, vehicles, and apparatus are properly made before fueling (or defueling) operations, and that the connections are not disturbed until after the completion of the operations.

5-3.2 Shipboard Fueling Precautions

Although the total elimination of all EMR arcing hazards to fuel may not be achievable aboard ship without placing unacceptable restrictions on flight operations and ship missions, the following practices are recommended to minimize the risk of accidental ignition:

a. Never energize transmitters on aircraft or vehicles in the vicinity of fueling operations.
b. Never make or break any electrical, static ground wire, or tiedown connection, or any other metallic connection, to aircraft, vehicles, or apparatus during fueling operations. Make connections before; break them afterwards.
c. Turn off all radars capable of mainbeam illumination of fueling areas, or inhibit them from irradiating the area by use of radiation cutout devices during fueling operations.
d. Do not energize HF transmitting antennas within the quadrant of the ship in which fueling operations are being conducted.

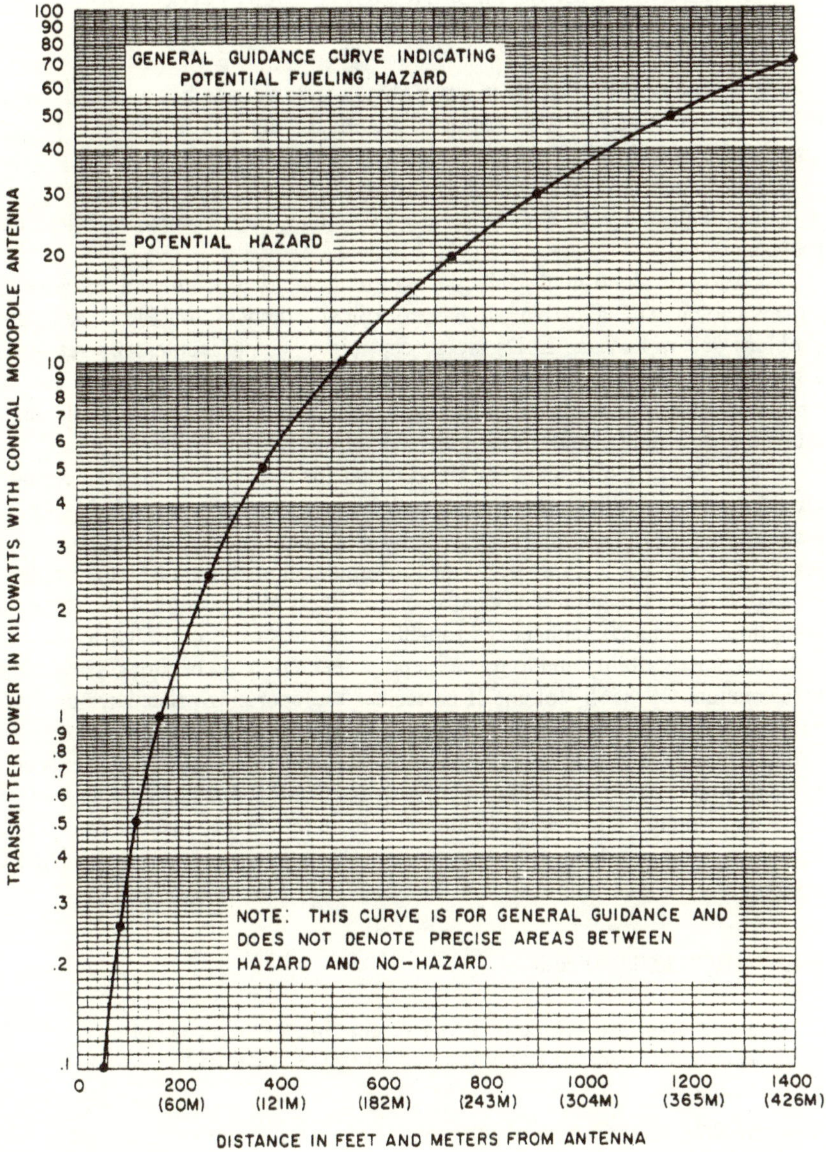

Figure 5-8 General Guidance Curve Indicating Potential Fueling Hazards

5-4 HAZARDS OF ELECTROMAGNETIC RADIATION TO ORDNANCE (HERO)

High-power EMR fields in naval ships create yet another potential hazard due to the sensitivity of some forms of electrically actuated explosives, propellants, and pyrotechnics. The hazard exists on virtually all warships but is perhaps most worrisome on aircraft carriers because of the necessity to frequently arm and disarm planes with a wide variety of ammunition, bombs, missiles, and rockets.

HERO results from the inherent nature of electrically initiated firing mechanisms known in naval parlance as electroexplosive devices, or EEDs. The HERO problem occurs because EEDs are susceptible to being accidentally set off or having their reliability degraded, by exposure to RF environments. The susceptibility has been found to be most critical during ordnance handling, loading, unloading, assembly, and disassembly operations.

Because of the concern for HERO, and for the safety of personnel under all shipboard conditions, the Navy has for many years sponsored an extensive testing program to determine the susceptibility levels of its ordnance to various forms of EMR. The tests are performed in simulated maximum RF environments which the ordnance and ordnance systems are likely to encounter, from stockpile conditions to launch sequence. From the tests, data are collected to classify ordnance susceptibility and to recommend proper safety precautions. Navy technical manual OP 3565 Volume II prescribes the operating procedures and precautions necessary for the safe handling, transporting, and storage of ordnance, and to prevent the premature initiation of EEDs in all situations in which exposure to EMR may exist.[17] The following discussion is a generalized summary of OP 3565 philosophy. The reader should refer to the technical manual for specific details regarding ordnance types, exposure limits, and minimum safe HERO distances.

5-4.1 HERO Classifications

Based upon the degree of EMR susceptibility three categories of HERO have been established: HERO SAFE, HERO SUSCEPTIBLE, and HERO UNSAFE ordnance.

a. *HERO SAFE Ordnance*—Items of ordnance that are sufficiently shielded or protected so as to be negligibly susceptible to EMR effects and that require no special RF environmental restrictions.
b. *HERO SUSCEPTIBLE Ordnance*—Items of ordnance that are moderately susceptible to EMR effects and require moderate RF environmental restrictions to preclude jeopardizing safety or reliability.

c. *HERO UNSAFE Ordnance*—Items of ordnance that are highly susceptible to EMR effects and require severe restrictions for some or all phases of employment. It is to be stressed that assembly or disassembly of ordnance, or subjecting ordnance items to unauthorized conditions and operations, can cause HERO SAFE ordnance to become HERO UNSAFE.

5-4.2 HERO Controls in Port and Territorial Seas

Several agreements have been reached between the United States and other nations with respect to preventing HERO accidents when ships are visiting ports or steaming in territorial seas; e.g.:

a. All operations involving HERO SUSCEPTIBLE and HERO UNSAFE ordnance must be curtailed while in port or in territorial seas.
b. While sailing territorial seas, a ship must maintain a distance of 1,000 yards from shore-based radio and radar transmitters and from radio and radar transmitters on oil or gas drilling rigs. Should it become imperative to go in closer than 1,000 yards, only HERO SAFE ordnance may be exposed.
c. While visiting foreign ports, ammunition and EEDs that are HERO SUSCEPTIBLE or HERO UNSAFE must be protected at all times from exposure to EMR, either by stowage below decks in metal ships or by stowage in shielded closed containers.
d. Where stricter national regulations than those above exist, the stricter regulations must be adhered to.

5-4.3 Shipboard HERO Controls

Through many years of experience and tests, the following general guidelines have been developed to reduce the risk of HERO:

a. During the time that an aircraft is being armed or disarmed, its radio and radar equipment must be turned off. If there are other aircraft in the vicinity of the loading area that are capable of radiating hazardous EMR fields, it must be ensured that these aircraft do not transmit RF energy within safe HERO separation distances. If transmitter equipment in the loading area must be operated for maintenance purposes, it must be ensured that the transmitter is connected to a dummy load antenna.
b. A separation of at least 10 feet must be maintained between any shipboard transmitting antenna and all ordnance, including HERO SAFE ordnance. For HERO SUSCEPTIBLE and HERO UNSAFE items, greater separation distances are required (see [17] for specific criteria). The safe separation

zones apply not only to the ordnance item itself but to any mechanical structure or object to which the ordnance is attached, such as a gun mount, or aircraft, or a missile launcher. There are, however, three exceptions which do allow the collocation of shipboard transmitting antennas, ordnance items, and ordnance systems within distances less than 10 feet:

1. When, regardless of frequency, an antenna is radiating less than five watts average power, then HERO SAFE ordnance may be located up to five feet from that antenna.
2. When an antenna is radiating two watts or less average power at frequencies greater than 100 MHz, then both HERO SAFE and HERO SUSCEPTIBLE ordnance may be located up to five feet from the antenna.
3. When all loading procedures have been completed, an aircraft with HERO SAFE ordnance may be parked up to five feet from the vertical projection of a lowered deckedge transmitting antenna. During actual loading operations, however, the aircraft must be no closer than ten feet from the vertical projection of the lowered antenna as shown in Figure 5-9.

Figure 5-9 Example of HERO SAFE Distances on Aircraft Carrier

c. All ordnance operations must be planned so that there is a minimum exposure of EEDs to the RF environment. Internal wiring and firing circuits must never be exposed to the RF environment by assembly or disassembly. All HERO UNSAFE ordnance must be transported in completely enclosed metal containers wherever possible. Igniters, primers, detonators, and other items containing EEDs such as electrically fired rocket engines, guided missile motors, and electronic or electrical fuzes must never be stored together in the same compartment or magazine within five feet of RF cables, waveguides, or any other radiating or transmitting equipment. Moreover, these items should be stored in metal containers.

d. Electrical contacts, electrode primers, and contact pins must not be allowed to touch any object capable of conducting RF energy during ordnance handling and loading operations. Objects capable of conducting RF energy include aircraft structures, bomb rack breeches, cartridges, and tools. Electrical connections to air-launched ordnance systems must not be made before the ordnance is racked to the aircraft. Electrical connectors to ordnance systems are the most likely paths for RF energy to enter. Racking an ordnance item to the aircraft first and tightening the sway braces before making electrical connections reduces the amount of RF energy induced into the internal circuitry of ordnance items. Umbilical cords and cable connections should be handled only when absolutely necessary. All open electrical connectors on ordnance must be covered with nonshorting caps to prevent the pins of these connectors from being touched accidentally. The caps should be removed just prior to connector mating and reinstalled promptly upon disconnection.

e. When handling ordnance in the vicinity of HF transmitting antennas during dockside replenishment, all loading hooks and metal steering lines must be insulated from cranes, booms, and wires by the use of nonconductive rope or insulators. During connected replenishment (CONREP) when physical contact between the ships has been made with metal cables, ship HF transmitters must not be permitted to transmit energy while HERO SUSCEPTIBLE or HERO UNSAFE ordnance is present on any weather deck. To ensure HERO safety during CONREP ordnance operations, both ships must operate under emission control (EMCON) conditions.

f. It is possible that, when conducting vertical replenishment (VERTREP) while under way, helicopters may fly through high intensity mainbeams of radars. If HERO SAFE ordnance is being transferred, a 50-foot separation must be maintained between the ordnance and any radiating antenna. If the ordnance is classified as either HERO SUSCEPTIBLE or HERO UNSAFE, but is enclosed within an all-metal container, it can be considered HERO SAFE during VERTREP transfer. HERO SUSCEPTIBLE ordnance may in some cases be transferred outside of containers as long as minimum safe HERO distances are maintained (see [17]).

g. During flight deck operations HERO UNSAFE ordnance must not be permitted on the flight deck unless appropriate EMCON conditions are invoked. All aircraft radio and radar transmitters must be off while the plane is being loaded or unloaded. If other aircraft in the loading area are capable of radiating hazardous RF fields they must be prohibited from transmitting energy, or, if energizing is imperative for maintenance reasons, the equipment must transmit into dummy load antennas. It must be ensured that no RF fields exceed the maximum allowable environment for HERO.

h. HERO UNSAFE ordnance is not permitted on hangar decks (whether hangar doors are opened or closed) unless appropriate EMCON conditions are invoked. EMCON restrictions on HERO SUSCEPTIBLE ordnance in hangar bays are the same as those imposed on flight decks for HF transmitters. However, operation of aircraft transmitters into dummy load antennas is permitted. During CONREP, when physical contact has been made between ships by using metal cables which extend into the hangar bay, unrestricted operations on HERO SUSCEPTIBLE ordnance is not permitted on the hangar deck.

i. Because of the extensive amount of high-power communications equipment installed on major command ships and on communications relay ships, unique HERO problems can arise when these ships approach within 24,000 feet of other naval vessels. Consequently, the following precautions must be observed:

   1. When operating within 24,000 feet of other ships, HERO requirements must be coordinated with those ships to confirm that no HERO UNSAFE ordnance is present on weather decks or hangar decks; otherwise, EMCON conditions are to be invoked.
   2. When within 3,000 feet of another ship, HERO EMCON conditions are required.

j. Radars operating at frequencies greater than 1.0 GHz should be prevented from directly illuminating ordnance or any metallic object or structure attached to the ordnance when within the minimum safe HERO distances. If HERO SUSCEPTIBLE ordnance will be in the mainbeam of a radar and inside the minimum safe HERO separation distance, the radar must be shut down. Radars operating at frequencies less than 1.0 GHz must be turned off whenever susceptible ordnance will be within the minimum safe HERO separation distance. For the case of communications equipment radiation fields, the safe distance field strengths for HERO SUSCEPTIBLE ordnance can be determined from Figure 5-10.

k. HERO UNSAFE ordnance can be protected from EMR by placing it in a completely enclosed all-metal container. When exposure of HERO UNSAFE ordnance cannot be avoided, it should be exposed only below decks

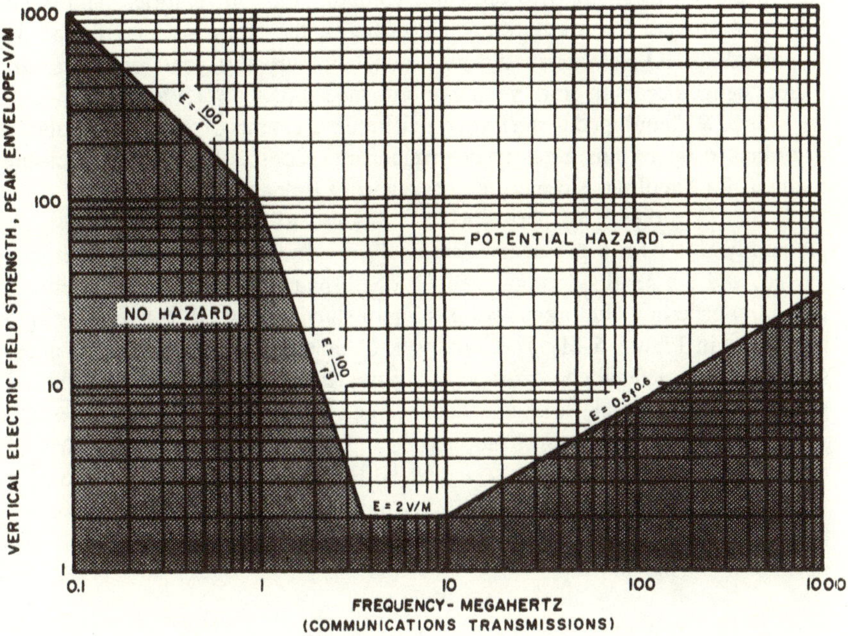

Figure 5-10 Maximum Safe Field Strength for HERO SUSCEPTIBLE Ordnance

in an RF-safe area. It must never be permitted on weather decks unless appropriate EMCON conditions are invoked. HERO SAFE and HERO UNSAFE ordnance can be classified as HERO UNSAFE by the following:

1. Assembling or disassembling of ordnance systems undergoing repairs, upkeep, or parts exchange.
2. Testing, such as resistance of continuity checks, using electrical connections to ordnance items.
3. Exposing unshielded or unfiltered wire leads of primers, blasting caps, impulse cartridges, and other EEDs.
4. Exposing unshielded ordnance subassemblies such as rocket motors, warheads, exercise heads, and fuzes.

5-4.4 Shipboard HERO Surveys

The EME of a ship changes with new or modified radar, EW, HF communication, and navigation transmitter installations. The environment also changes significantly with changes to ordnance configurations, inventories, and operations. Because of these environmental changes, the Navy has determined that shipboard HERO surveys should be conducted every five years or whenever a major change occurs in emitter suite or ordnance allocation. HERO survey teams are trained and equipped to perform on-site measurements of the RF environment in ordnance operations areas to determine the specific HERO safety measures required for handling, storage, and transport of ordnance items.

HERO surveys are performed in response to requests from ships. The process begins with a presurvey data analysis. An on-site survey is then conducted, and the results are analyzed for conformance to established safety and reliability criteria including proper posting of standardized HERO warning signs illustrated in Figure 5-11. The survey is completed with the preparation and submission of a detailed report which provides the survey findings, analysis results, conclusions, and recommendations. This report becomes the single source of ship-specific technical data to support the individual shipboard HERO EMCON restriction directives; i.e., the so-called HERO EMCON BILL. Therefore, by performing measurements of the EME in a most-to-least order of hazard potential, the overall results are assessed to relate best to the current and future safe ordnance operations for the ship.

Figure 5-11 HERO Warning Symbol

REFERENCES

1. K. R. Foster and A. W. Guy, "The Microwave Problem," *Scientific American*, Vol. 255, No. 3, September 1986, p. 32.
2. A. W. Laine, "Electromagnetic Radiation in Prolonged Space Environments," *Interference Technology Engineers Master*, Robar Industries, West Conshohocken, PA, 1985, p. 152.
3. E. J. Lerner, "Biological Effects of Electromagnetic Fields," *IEEE Spectrum*, Vol. 21, No. 5, May 1984, p. 57.
4. *Radiation Hazards Handbook*, Department of Defense ECAC-HDBK-86-005, Electromagnetic Compatibility Analysis Center, Annapolis, December 1986, p. 2–1.
5. Lerner, *op cit.*, pp. 59–60.
6. M. G. Morgan et al., "Powerline Fields and Human Health," *IEEE Spectrum*, Vol. 22, No. 2, February 1985, p. 62.
7. L. E. Polisky, "The Commercial Application of Non-Ionizing Electromagnetic Radiation Hazard Level Standards," *EMC Expo 86 Symposium Record*, International Conference on Electromagnetic Compatibility, Washington, DC, June 1986, p. T24.4.
8. Lerner, *op. cit.* pp. 58–59.
9. Foster and Guy, *op. cit.* p. 32.
10. *Ibid.*, p. 33.
11. "Personnel Protection Policy for Exposure to Radio-Frequency Radiation," OPNAVNOTE 5100, Department of the Navy, Washington, DC, 30 July 1985.
12. N. T. Baron, "A New Radio-Frequency Radiation Criteria for the U.S. Navy," *EMC Expo 86 Symposium Record*, International Conference on Electromagnetic Compatibility, Washington, DC, June 1986, p. T24.11.
13. M. Z. Netzer, "Erroneous Measurements of Stray Magnetic Radiation," *Interference Technology Engineers Master*, Robar Industries, West Conshohocken, PA, 1986, p. 226.
14. *Electromagnetic Radiation Hazards to Personnel, Fuel, and Other Flammable Material*, Technical Manual NAVSEA OP 3565, Vol. I, Department of the Navy, Washington, DC, 15 July 1982.
15. *Ibid.*, p. 3–2.
16. *Ibid.*, p. 7–1.
17. *Electromagnetic Radiation Hazards to Ordnance*, Technical Manual NAVSEA OP 3565, Vol. II, Department of the Navy, Washington, DC, 1 August 1986.

Chapter 6
Shipboard Electromagnetic Pulse (EMP)

6-0 PREPARATION FOR AN EVENTUALITY

We now turn our attention to a most unusual electromagnetic phenomenon, one that is of extreme concern to shipboard electronics but which actually has never been experienced by naval ships except in simulated low-level testing. The phenomenon is electromagnetic pulse, or EMP. So high is the potential for harm done by EMP that one news columnist, while acknowledging that it "is still no more than a scientific theory mercifully untested," has described it as "awesome," and a "forbidding new destructive force." The columnist went on to say:

> But what EMP means to the rest of us is simply this: If nuclear weapons were to be detonated 200 miles above the United States, the electromagnetic pulses from the explosion would almost instantaneously knock out all the electrical power in North America. No television, no radio, lighting, hospital equipment, computers, telephones. Total blackout of the entire continent.
>
> . . . What worries our [military] strategic thinkers, though, is that EMP might be used to knock out America's top level C^3 system—command, control, and communications—that is supposed to respond to a nuclear strike with a retaliatory attack.[1]

The news report quoted above is alarming, and, of course, was written in a manner precisely to raise alarm. It is not however, overstated. The analysis is accurate and the concern is genuine—for military, civil, and commercial interests.

The potential for widespread disruptive effects resulting from EMP has been known for more than 20 years. In fact, one of the first public reports appeared in the autumn of 1967, where, in an electronics trade journal, it was noted that:

> During the high altitude nuclear tests in the Pacific in the early 1960s, ''hundreds of burglar alarms'' in Honolulu began ringing. ''Circuit breakers on the power lines started blowing like popcorn.''[2]

Because there were no electrical storms anywhere in the vicinity it was soon determined that intense electromagnetic energy radiated from a high-altitude atomic test 800 miles from Hawaii had created the unusual disturbances.

Scientists conducting the tests were aware of strong electromagnetic effects while observing the overload of sensitive measurement instruments and the upset of communication links. It is only in our modern era of more sophisticated means of deploying and detonating high-yield nuclear devices so as to cause massive, deliberate upset of delicately vulnerable solid-state electronic systems, however, that EMP has been recognized correctly as a ''forbidding new destructive force.'' One has only to imagine the chaos that would result from electrical shutdown of the highly computerized commercial sector of our society in banking, telecommunication, power utilities, stock exchange, mass transportation networks, and medical facilities, all from some unseen, unannounced, mysterious electromagnetic force from a far-off, otherwise harmless nuclear explosion.

A threat of such severity and magnitude cannot be lightly regarded. It has prompted much study and analysis, especially over the last decade. We hope that no society will ever have to experience EMP from a nuclear weapon explosion. Nevertheless, so long as we endure in an imperfect world we must be fully prepared for the eventuality. Indeed, techniques to harden electronic systems (and, for our purpose, ships) against the effects of EMP are being devised and implemented.

6-1 EMP CHARACTERISTICS

It is important to be clear about what we mean by EMP. The generation of electromagnetic pulses is, in the broadest sense, a routine occurrence in many ordinary types of electronic systems. A familiar example is the use of radar transmitters to produce narrowband pulsed electromagnetic energy which is purposely radiated outward to search for and track selected targets. The term EMP as generally accepted in the engineering community, however, is not the gentle pulses of energy created in myriad fashion by electronic circuitry and systems, no matter how complex or high in power level or short in duration. Rather, EMP is widely understood to mean an extremely intense, highly threatening, instantaneous, wideband pulse of electromagnetic energy originating from a fearful source: a nuclear explosion. To leave no room for doubt of its origination, some scientists and engineers prefer the more precise term nuclear electromagnetic pulse (NEMP). At the present time, however, EMP is still the more commonly used and recognized short form. Therefore, it will be employed exclusively hereafter in our discussion.

As depicted in Figure 6-1, there are four basic regions in which electrical and electronic systems may be subjected to the effects of EMP: at or near ground level, in the lower atmosphere, in the upper atmosphere, and at exoatmospheric altitudes. Since our particular interest is in what might happen to shipboard systems, our attention is focused on the effects at ground level.

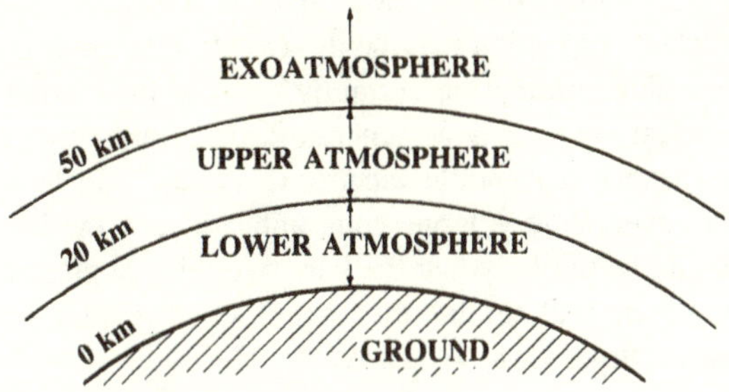

Figure 6-1 System Operating Categories

Going a step further, nuclear explosions may be similarly classified as one of three types: surface, air, or exoatmospheric. Surface and near surface bursts occur nominally at heights of ground zero to about 2 kilometers. Air bursts take place between approximately 2 and 30 kilometers, and exoatmospheric explosions are those which happen above 30 kilometers. Exoatmospheric detonations are frequently referred to as high-altitude EMP, or HEMP.

Damage caused by nuclear explosions is a function of weapon size (i.e., yield) and proximity to vulnerable systems. The principal burst effects are blast, heat, shock, and ionizing radiation of neutrons, x rays, and gamma rays. Should the burst occur near the earth's surface or in the low atmosphere in the general vicinity of a ship, the physical damage would be overwhelming, resulting in local devastation beyond the scope of our interest in the effects of EMP. Consequently, it is nuclear detonation in the exoatmospheric region that is of concern to us. It must be assumed that exoatmospheric nuclear bursts are a favored weapon option as they have the potential for dramatically affecting electrical and electronic systems from a very great distance, severely disrupting these systems without doing a pinch of other damage; i.e., in the absence of any of the other nuclear effects such as shock, heat, blast, or ionized radiation.

6-1.1 High-Altitude EMP Generation

Figure 6-2 is an artist's conception of a nuclear explosion occurring high above a naval fleet. Note, however, that such an explosion should not be perceived as always the result of an enemy attack. It could happen as well from detonation of one of our own, or an ally's defensive weapons; it could be from a nuclear engagement between third-party nations, or, conceivably, even from a nonaggressive high-altitude test in violation of current test ban treaties. The resultant effects on unprotected electronic systems nevertheless would be the same, irrespective of the reason for initiation of the burst. As a matter of fact, it is one of the subtleties of EMP that the immediate reason for and the location of a nuclear detonation may be difficult to discern or predict accurately. Yet it is a reasonable assumption certainly that the motivation for exploding a high-altitude nuclear weapon is to generate a pulse of energy of such intensity as to upset or disable susceptible electronic systems, including those aboard naval warships, over a very large geographic area.[3] In the shipboard case, moreover, it would be unlikely that the burst would take place directly overhead (as suggested in Figure 6-2) because the same destructive effects could be achieved if the explosion occurred from far away.

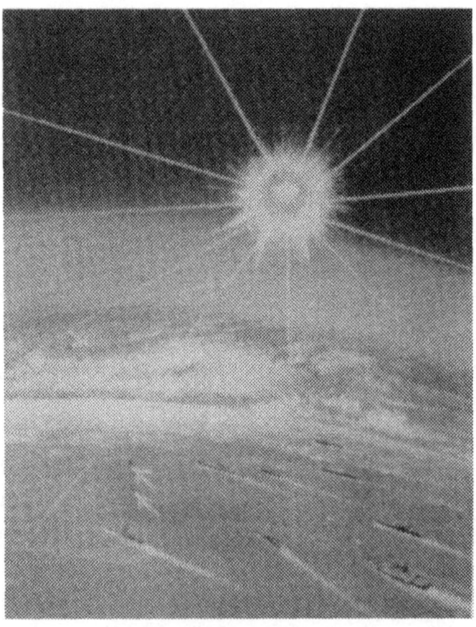

Figure 6-2 Conceptual Illustration of Nuclear Burst

Upon explosion at high altitude, all the emitted nuclear products spew radially outward from the burst center. Most are dissipated in the thin exoatmospheric medium and outer space. Those directed toward the earth, however, quickly encounter the lower atmospheric regions where the remaining products, except for EMP originators, are effectively absorbed. Figure 6-3 illustrates this event. When gamma rays from the explosion meet the atmosphere they interact in such a manner as to create electromagnetic energy in a process of physics known as the Compton Effect. By this process the newly generated energy is propagated as an electromagnetic field over great distances from the source.

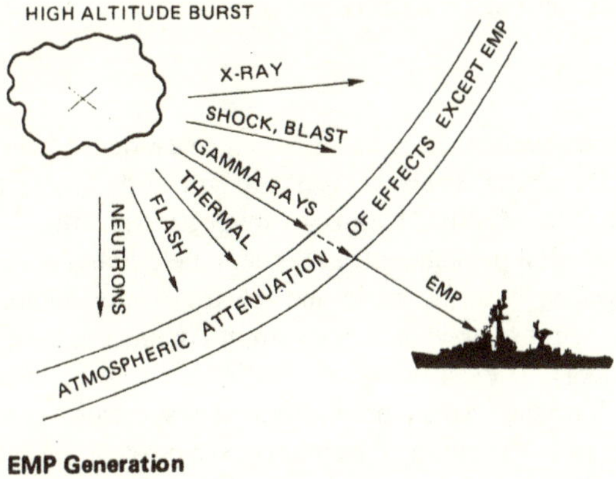

Figure 6-3 EMP Generation

The Compton Effect, essential to the creation of EMP, is described as follows:[4]

Gamma rays (and, to a much lesser degree, x rays) emanating as photon energy from the explosion reach the atmosphere and begin colliding with air molecules and dust particles. The collisions are of such force as to dislodge and scatter electrons from the molecules. The ejected electrons, now known as Compton electrons, are accelerated predominately in the former direction of the gamma rays; i.e., toward the earth's surface, as pictured in Figure 6-4. This process of separation of charge produces an electric field, and the electron movement constitutes an electric current, with an associated magnetic field. However, the process has not yet created classic electromagnetic radiation.

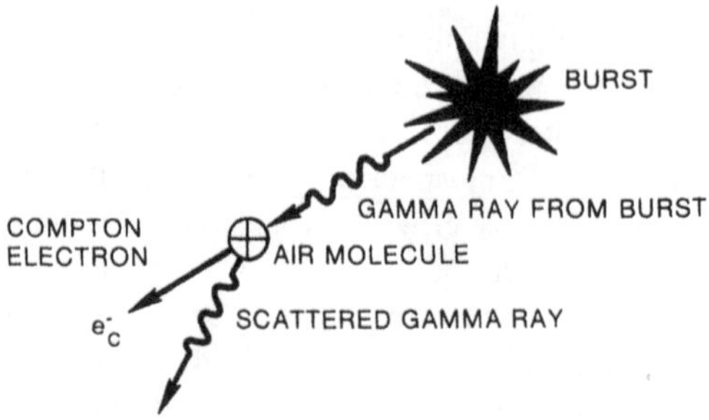

Figure 6-4 Compton Scattering Process

The chief mechanism which acts to produce a radiated field is the deflection and twisting of the Compton electrons as a result of the interactive force of the earth's magnetic field. Modified by this geomagnetic field, the Compton electrons begin to follow a spiral path about the magnetic field lines, as depicted in Figure 6-5. Now possessing both magnetic and electric vector components that vary with time, the electrons, moving as a coherent composite, have been efficiently converted in energy to electromagnetic radiation. The radiated fields are extremely high in intensity, have a broad frequency spectrum, and, because of the height and extent of deposition, instantaneously cover a very large area of the earth's surface. Because of the highly specialized nature of the radiated field it is quite properly characterized as EMP.

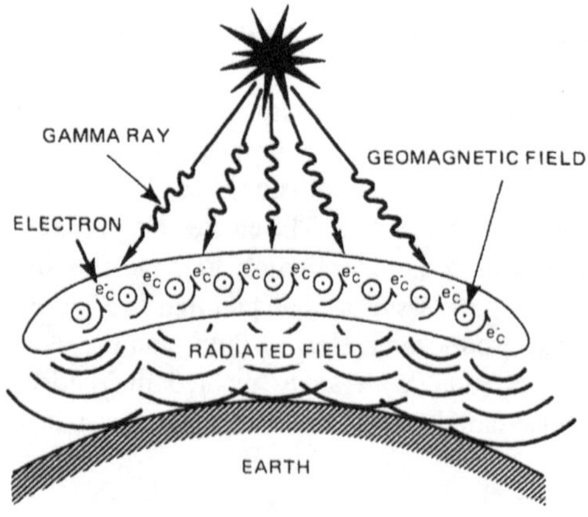

Figure 6-5 EMP Radiation Field Generation

6-1.2 High-Altitude EMP Electrical Properties

The far-reaching consequences of a high-altitude nuclear explosion are immediately apparent from Figure 6-6. If a 1-megaton nuclear bomb were detonated at approximately 300 miles above the center of the United States, the entire nation would suffer the effects of EMP with little or no other indication that a nuclear burst had occurred. Likewise, if the explosion happened over a large body of water such as the Indian Ocean or Mediterranean Sea, all ships within a very large area would be affected.

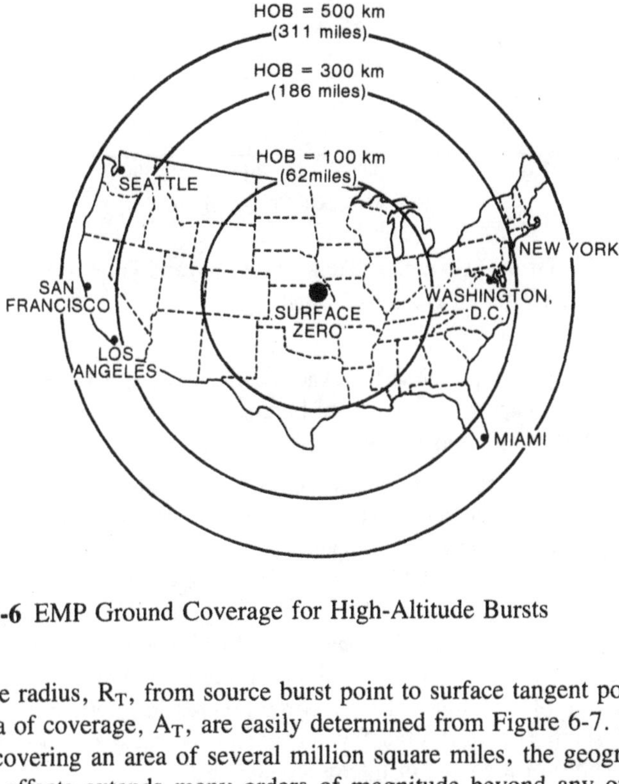

Figure 6-6 EMP Ground Coverage for High-Altitude Bursts

The radius, R_T, from source burst point to surface tangent point, and the total area of coverage, A_T, are easily determined from Figure 6-7. It is evident that by covering an area of several million square miles, the geographic range of EMP effects extends many orders of magnitude beyond any other nuclear effects. This is the major reason that exoatmospheric explosions must be anticipated. But equally important is the nature of the pulse itself. Although sometimes likened to the energy in a lightning stroke, EMP is actually quite different from any other natural or man-made electric phenomenon. The spectrum for EMP is broadband, extending from extremely low frequencies to very high frequencies, and the pulse has a much higher amplitude and faster rise time than, for example,

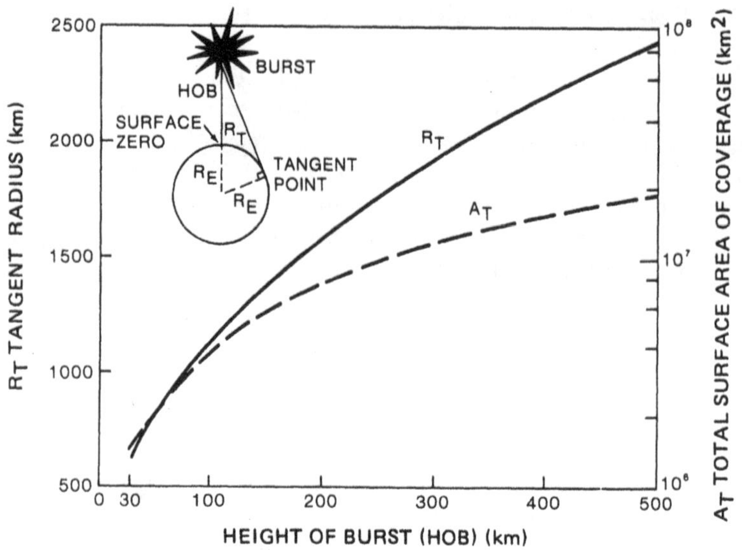

Figure 6-7 EMP Ground Coverage (Tangent Radius) and Total Area of Coverage as Functions of Height of Burst

a nearby bolt of lightning. While the exact characteristics of EMP are complex and depend upon weapon size, height of burst, and atmospheric conditions, the following properties are considered representative:[5]

a. *Field Intensity*—Based on free space impedance calculations, EMP energy can reach a peak field strength of up to 100 kilovolts per meter with H-field intensities of over 250 amperes per meter.
b. *Frequency Spectrum*—EMP occupies a broad bandwidth with damaging effects from 10 kHz to 100 MHz and peak intensities between 1 and 10 MHz. As such, the spectral content of EMP incorporates the frequencies used by a great many commercial and military electronic systems.
c. *Waveform*—High-altitude EMP, as represented in Figure 6-8 (from [3]), has a sharp rise time of a few nanoseconds and a duration (effective pulsewidth) of a couple of microseconds.
d. *Polarization*—EMP generated from a high-altitude nuclear explosion is propagated downward from the source region in a horizontally polarized plane wave. Local polarization depends on latitude and longitude of the burst and relative location of the sensor. Therefore, EMP energy is eminently suited for interception and collection by large vertical and horizontal bodies of metal, such as a ship hull, and many metallic items on the hull like masts, lifelines, fan antennas, cables, waveguides, pipes, and ducts.

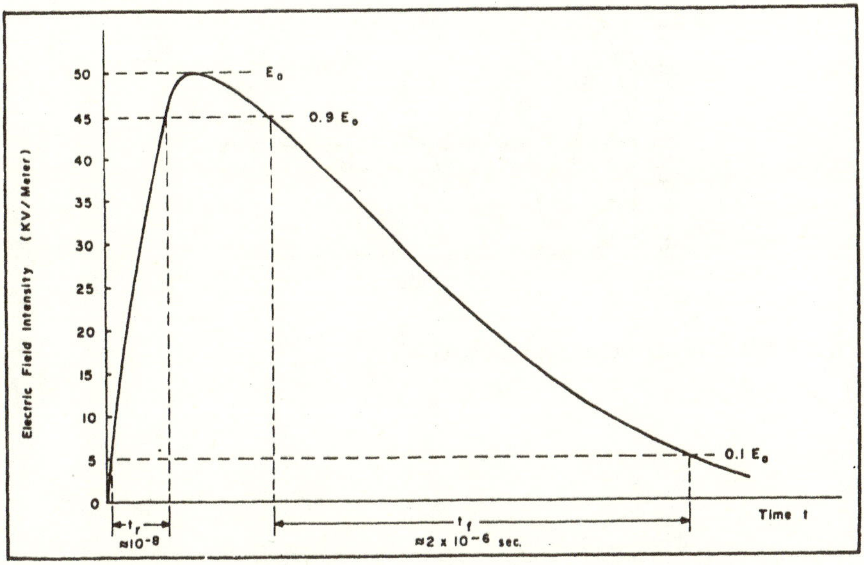

Figure 6-8 Representative EMP Waveform for High-Altitude Burst

From these unique characteristics it can be appreciated that EMP radiation, although brief in existence, is billions of times more intense than an ordinary radio signal.

6-2 SHIPBOARD EMP DAMAGE EFFECTS

Metallic objects exposed to an electromagnetic field will serve as receptors of radiated energy. That is, they will act as a rudimentary form of receiving antenna even though they are never intended for that purpose. Generally, the larger the metallic structure, the greater the amount of collected EMP energy. Naval ships, obviously, are very large metallic structures. When EMP impinges upon a ship, some of the energy penetrates directly to below-deck compartments through hatches, doorways, windows, hull gaps, and seams. Most of the received pulse, however, is transferred to interior electronic systems by shipboard antennas (via associated transmission lines and waveguides), external cables, pipes, ducts, and conduits, whence it couples to wiring, cabling, and equipment appendages or passes through enclosure apertures and poorly shielded barriers in equipment to inflict a sudden surge of high ringing current like that of Figure 6-9 on sensitive electronic circuits. Figure 6-10 symbolically portrays some of the many paths by which EMP can gain access to interior equipment in a ship.

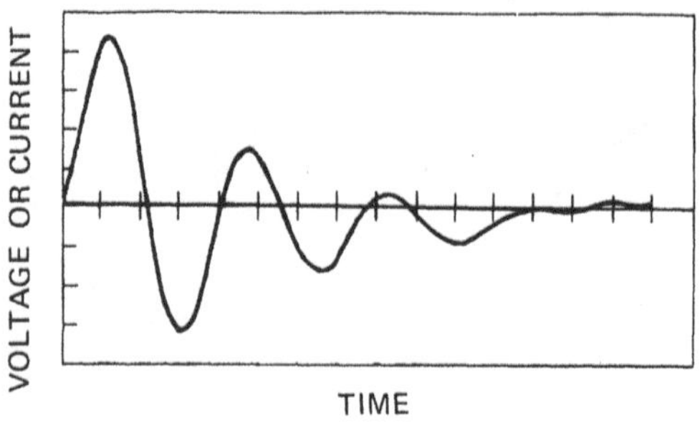

Figure 6-9 Ringing Response Characteristics

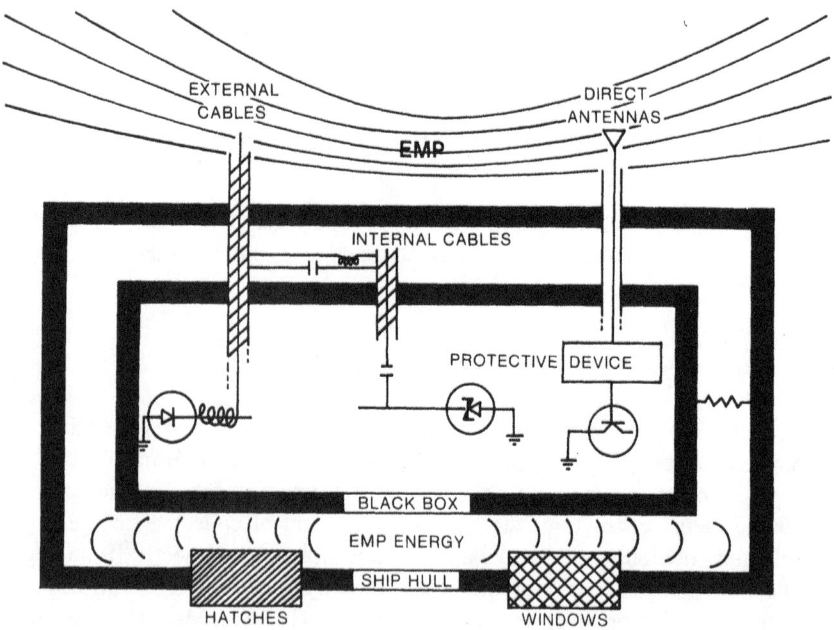

Figure 6-10 EMP Coupling and Penetration

Electrical and electronic systems are disrupted by EMP in one of two fundamental ways: either through physical damage of discrete components or by circuit upset. In the case of component damage, the usual effect is burnout of microminiature solid-state devices or other forms of electrical degradation of such severity that an element in the circuit no longer operates within its design parameters. Circuit upset, on the other hand, normally results in loss of data transmission or loss of stored memory. Upset is far more likely an event than physical damage because the energy required to upset a circuit is at least an order of magnitude less than that required to burn out a component.[6] Upset of a system occurs when an induced EMP within the circuit time response exceeds the operational level. Appearing as a false transient signal, the EMP can change the state of a logic element, cause loss of clock and synchronization, or erase memory. Disruption of operations can be so severe, especially in the event of stored memory loss, that reprogramming and reloading of data may be required. As a consequence there may be long periods of system outage. Such disruption to mission-critical operations could seriously impair a warship's capability to fight, and is, therefore, intolerable. EMP, although a short-term impulse, can have disastrous long-term effects.

Circuit components most likely to be adversely affected by EMP are those with low power ratings and critical operating characteristics where small changes produce significant effects. Semiconductors are prime suspects, and since they are so vulnerable if left unprotected, it is well to examine the reasons they fail. The preponderant cause of semiconductor failure is thermal overload, which results in junction melt and a short circuit. Burnout of this type generally happens when the EMP imposes a sudden reverse bias on the junction to drive it into breakdown. Failure can result from forward stressing of a junction, too, but the forward-direction threshold is several times higher because of the low impedance and voltage tolerance offered in forward conduction.

Other electronic components are susceptible to EMP disabling to a much lower degree. Resistors can change value when overheated by high pulse power. Capacitors can suffer dielectric breakdown from excessive transient voltage, and such elements as switches, relays, coils, and transformers may experience insulation flashover, arcing at contacts, and melting of wiring. The EMP voltage spike may initiate a momentary breakdown path that, once established, is sustained by normal circuit operating levels.[7] Laboratory experiments have demonstrated that the old electronic vacuum tube circuits were more resistant to damage from EMP effects than are semiconductor systems. The transformation of vacuum tube circuits, which were relatively hard to EMP, to delicate transistors and integrated circuit components, was of course never anticipated during the atomic test periods of a quarter-century ago. This transition to microminiature, sensitive, low-power, solid-state electronics has resulted in the dramatic increase in EMP vulnerability.*

6-3 SHIPBOARD EMP HARDENING TECHNIQUES

The subject of EMP was purposely discussed near the end of this book because the methods used to protect and suppress the effects of EMP encompass most of the practices and philosophy discussed previously, e.g., enclosure shielding, cable shielding, bonding, grounding, isolating (decoupling), and compatibility with the shipboard EME. The methods formerly examined do not necessarily all have direct application to EMP mitigation because of the unique and severe nature of EMP; nevertheless, EMP hardening techniques are in many respects evolutions of common EMI suppression practices.

By way of testimony to the seriousness of EMP, Pinkston [5] has noted that, although there is growing concern by the commercial electronics community and the public services over vulnerability of their systems, it is the armed forces that have responded to the potential threat with immediate action: "EMP is the most consistently specified environment in the nuclear hardening of military electronic equipment." The main thrust of this action is to provide adequate protection. It is imperative that the nation's defense systems be sufficiently hardened against failure caused by such events as logic circuit upset of missile guidance control or interruption of crucial command and control coordination by the burnout of input stages of, say, shipboard communication receivers.

The goal of shipboard EMP protection is to prevent the pulse energy from entering areas containing susceptible equipment and systems.[8] This requires effective shielding or isolation of the equipment from the external EME, and, at the same time, the use of less susceptible electrical and electronic systems. The all-metal construction of ships with thick steel plating and the technique of using continuous-weld seams would appear to provide a near-ideal EMP shield. But the many hull and superstructure penetrations required for normal ship functioning—the antenna transmission lines and cables, ducts, doorways, windows—degrade shielding effectiveness. The manner in which interior compartments of a ship are fashioned also affords a good degree of additional shielding; these interior spaces too must have openings and intrusions, however, which reduce the overall shielding integrity. The essence of providing adequate shipboard EMP protection, then, is properly to control or treat the many openings

*By definition, *susceptibility* is the ability of the system to detect the threat, and *vulnerability* refers to the inability to survive, given detection of the threat.

and penetrations in order to take advantage of the inherent quality of a ship's metal structure.

For purposes of EMP engineering, system resistance to EMP is classified as either hard or soft. Hard systems are those that are specifically designed to withstand the effects of a nuclear environment so as to continue functioning normally. Soft systems are those not designed to operate in a nuclear environment, so they must be protected by the enclosure in which they are contained. Insofar as possible soft systems should be made intrinsically less vulnerable (less collection of energy and less coupling efficiency) to EMP by the use of harder components. Otherwise, the only reasonable alternative is to keep EMP energy from reaching soft systems; i.e., to reduce susceptibility. General guidelines for minimizing EMP exposure include:

a. Shield the system within a metallic enclosure. Reduce to a minimum the number of apertures and aperture sizes. Bond all seams. Use RF gaskets on hatch covers and doors. Use wire mesh or transparent EMI film coatings over windows and viewports.
b. Route cables inside the ship structure, inside masts, and inside conduits to the maximum extent possible. Use as few and as short cables as possible. Employ tightly braided or continuous foil cable shields, terminated at the enclosure periphery with conductive backshells.
c. Eliminate ground loops if possible, or keep them at bare minimum by proper grounding practices.
d. Isolate sensitive internal electronics such as microprocessors and memory circuits.
e. Use nonconductive interface data lines such as fiber optics where practicable. Otherwise use highly shielded twisted pair lines and redundant data lines. Fiber-optic cables are immune to EMP coupling, so are preferred.
f. Choose least-sensitive electronic circuit components.
g. Use filters on interface lines that will withstand EMP transient energy.
h. Use terminal protection elements such as amplitude limiting devices and circuit breakers to shunt or disconnect pulse energy from sensitive circuitry.

EMP, as a threat to the overall ship mission, must be considered on a total system basis throughout all phases of design and operation. The two major engineering techniques used for EMP protection are cable shielding and use of circuit protection devices.

6-3.1 EMP Shielding and Grounding

One of the most effective methods of hardening a ship against the threat of EMP is to enforce proper shielding and grounding. Cables in particular must be prevented from picking up and transferring energy from exposed topside areas into the ship's internal compartments. Even the complete closing up of all hull apertures would prove futile if EMP transients were allowed to be conducted freely on cables that penetrate to the inside. If the cables are poorly shielded, the EMP energy will couple directly to the cable inner conductors and thence will flow to the ship interior, where it will be applied suddenly to the input of equipment to which the cables are connected. Furthermore, part of the energy will radiate from the cables to cause cross-coupling into other systems not even associated with the original exposed cable.

By far the best way to reduce the potential for collection of EMP energy is to shield all topside cables completely; i.e., house the cable conductors inside a metal shroud. Where possible this should be accomplished by restricting cable runs to the ship interior so as to take good advantage of the innate, though imperfect, shielding characteristics of the hull. For cables that must be routed outside, the use of solid metal conduit or trunks is recommended. Well-grounded conduits and trunks will act to intercept the incident EMP and disperse it harmlessly over the external skin of the ship. At all points where the conduit penetrates the hull, it must be welded circumferentially at the point of entry (e.g., deck and bulkheads) on the external side. Cables leaving the main deck must also be enshrouded in conduits as detailed in Figure 6-11.

To achieve sufficient reduction of the hundreds of RF amperes that may be induced on an outer cable shield from EMP, at least 80 dB of attenuation is needed.[9] The most practical way to keep this current from being applied to below-deck systems is to shunt the energy to the ship ground at each point where the cables traverse a bulkhead or a deck boundary from topside to interior.

6-3.1.1 Cable Shielding Requirements

Navy requirements specify that all cables routed in shipboard topside areas must be shielded from EMP. Coaxial cables and others having an overall inherent shield must have the shield grounded at deck or bulkhead penetration points to remove EMP energy from the cable prior to its passing to the interior. Cables with an overall solid shield are EMP-protected and require no further shielding. Cables exceeding these provisions of exposure, and all unshielded cables and wires, must be enclosed in a solid conduit pipe, in a flexible conduit, or in a metal trunk, with a cable shield grounded to the enclosure points of entry and exit as shown in Figure 6-12. For cable access, wireway trunks must have removable covers using captive bolts on both sides of the cover, with spacing not to exceed 12 inches to ensure proper metal-to-metal contact of the cover to the trunk. Any nonmetallic boxes or covers used for topside cable connections

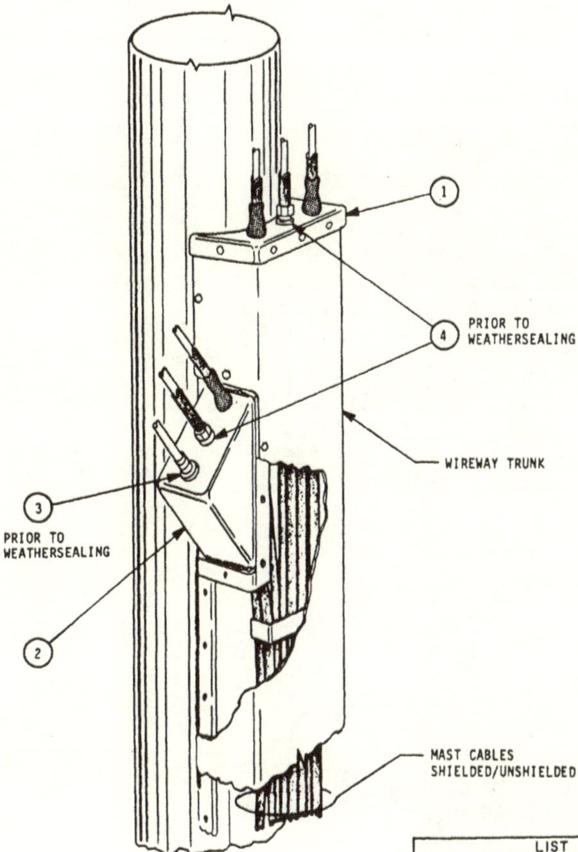

LIST OF MATERIAL			
ITEM NO.	PART	SPECIFICATION	NOTE
1	END CAP		1
2	BOX, BREAKOUT		1
3	ADAPTER, GROUNDING	COMMERCIAL	4
4	FITTING, CONDUIT	COMMERCIAL	4

NOTES:

1. FABRICATE END CAP AND BFEAK-OUT BOXES, AS REQUIRED, TO ACCOMODATE THE NUMBER OF AND TYPE OF CABLES THAT EXIT THE WIREWAY TRUNK. HOLES SHALL BE SIZED TO FIT THE REQUIRED GROUNDING ADAPTERS AND CONDUIT TERMINATION FITTINGS.

2. AFTER CABLE TYPES AND GROUNDING AND SHIELDING REQUIREMENTS HAVE BEEN PREDETERMINED, INSTALL THE CABLES THROUGH THE END CAPS, BREAKOUT BOXES, GROUNDING ADAPTERS AND THE CONDUIT FITTINGS.

3. MEASURE AND CUT THE REQUIRED LENGTHS OF SHIELDING CONDUIT TO SHIELD ALL REQUIRED CABLES.

4. INSTALL GROUNDING ADAPTERS AND FLEXIBLE SHIELDING CONDUIT. WEATHERSEAL AS SPECIFIED FOR CORROSION PROTECTION.

Figure 6-11 Mast Cables Located Within Wireway Trunk

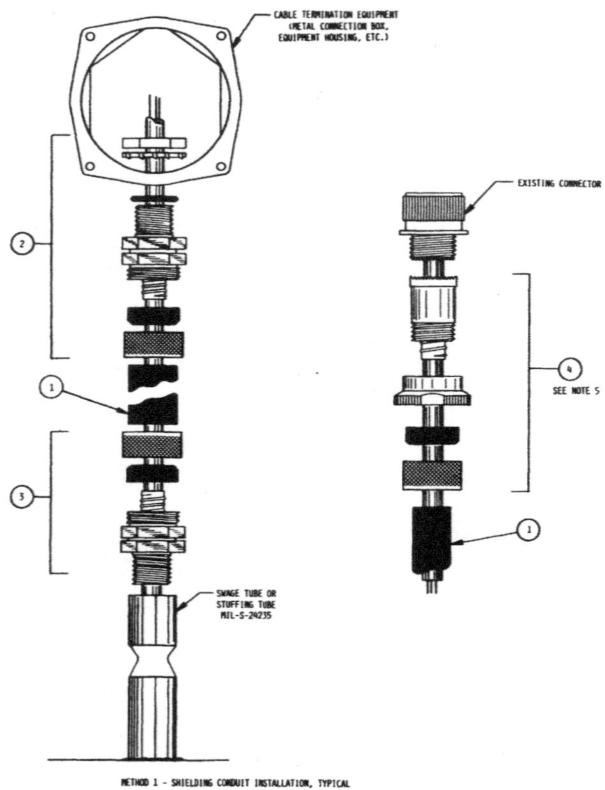

Figure 6-12 Cable Shielding Methods

LIST OF MATERIAL			
ITEM NO.	PART	SPECIFICATION	NOTE
1	CONDUIT, SHIELDING	COMMERCIAL	4,6
2	FITTING, CONDUIT-TO-BOX	COMMERCIAL	4,6
3	FITTING, CONDUIT-TO-STUFFING TUBE	COMMERCIAL	3,4,6
4	FITTING, CONDUIT-TO-CONNECTOR	COMMERCIAL	4,5,6

NOTES:

1. THIS METHOD OF CABLE SHIELDING APPLIES TO NEW CABLE INSTALLATIONS AND TO EXISTING CABLE INSTALLATIONS WHERE THE CABLE CAN BE REMOVED FROM THE TERMINATING EQUIPMENT AND HANGERS AND ROUTED THROUGH THE CONDUIT.

2. FOR NEW CABLE INSTALLATIONS - PRIOR TO PULLING CABLE THROUGH STUFFING TUBE, REMOVE PACKING GLAND NUT FROM THE TUBE AND DISCARD.

   FOR RETROFIT INSTALLATIONS - REMOVE CABLE FROM TERMINATING EQUIPMENT (LIGHT, CONNECTION BOX, SPEAKER, CONNECTOR, ETC.) AND REMOVE CABLE FROM HANGERS DOWN TO STUFFING TUBE. UNSCREW PACKING GLAND NUT FROM STUFFING TUBE, REMOVE NUT FROM CABLE AND DISCARD.

3. FOR BOTH NEW AND RETROFIT INSTALLATIONS - SELECT CONDUIT-TO-STUFFING TUBE END FITTING TO MATCH CONDUIT SIZE AND TUBE SIZE. ROUTE CABLE THROUGH FITTING. PACK STUFFING TUBE (NEW INSTALLATIONS). COAT LOWER THREADS OF FITTING WITH ANTI-SEIZE COMPOUND OF MIL-T-22361. SCREW FITTING INTO STUFFING TUBE AND TIGHTEN AS REQUIRED FOR PACKING.

4. MEASURE AND CUT PROPER LENGTH OF CONDUIT TO COVER THE ENTIRE LENGTH OF CABLE FROM STUFFING TUBE TO END TERMINATION. ENSURE CONDUIT IS CUT SQUARE. FEED CABLE THROUGH CONDUIT. COAT UPPER THREADS OF FITTING WITH ANTI-SEIZE COMPOUND AND TERMINATE CONDUIT INTO FITTING AT STUFFING TUBE. INSTALL CONDUIT IN CABLE HANGERS AND TERMINATE OTHER END OF CONDUIT INTO END FITTING AT TERMINATING EQUIPMENT. ENSURE ALL METAL CONNECTING PARTS ARE COATED WITH ANTI-SEIZE COMPOUND PRIOR TO ASSEMBLY. CONNECT OR RECONNECT CABLE CONDUCTORS TO PROPER TERMINALS.

5. WHERE BOTH ENDS OF THE CONDUIT TERMINATE IN AN ELECTRICAL OR ELECTRONIC CONNECTOR, THE METHOD OF CONDUIT ATTACHMENT TO CONNECTOR MAY BE IN ACCORDANCE WITH MIL-C-28840.

6. SHIELDING CONDUIT AND ASSOCIATED END FITTINGS SHALL BE THE FOLLOWING, OR EQUAL. "OR EQUAL" SHALL BE DEFINED AS EQUAL IN ELECTROMAGNETIC SHIELDING, GROUNDING, CORROSION PROTECTION, AND WEATHERPROOFING.

   BREEZE-ILLINOIS, INC., WYOMING, IL,"BI-PRO 175" (TYPE SHOWN ON THIS FIGURE)

   ANAMET, INC., WATERBURY, CT, "SHIELDTITE"

   ETCON CORP., PALOS HEIGHTS, IL, "TYPE CC"

   GLENAIR, INC., GLENDALE, CA, "SERIES 75"

7. WEATHERSEAL CONNECTORS AND FITTINGS AS SPECIFIED FOR CORROSION PROTECTION.

or fixtures must be replaced by metallic boxes and covers for proper grounding of conduit. The outer shield of solid overall shielded cables must use the procedures illustrated in Figure 6-13 to ensure correct grounding at weather penetration points.

Cables routed inside the ship structure must not be installed within 12 inches of weather doorways, hatchways, or windows, and must be at least 10 feet from hangar doorways, unless the cables are double-shielded or enclosed in conduit. Cables that terminate at hull openings, e.g., windshield wiper cables, window deicing cables, and door alarm cables, must be placed inside conduits.

6-3.1.2 Waveguides, Pipes, and Metal Tubes Grounding

All waveguide transmission lines, metal pipes, and metal tubes that transit from topside areas to interior spaces must be grounded at each point of penetration, using the methods of Figure 6-13 for pipes and metal tubes and of Figure 6-14 for waveguides. Pipes, to be considered properly grounded, must be welded 360° circumferentially at penetration points or be threaded with fittings which are welded at penetration points.

6-3.2 Circuit Protection Devices

Of all the many possible paths for EMP to be conducted into sensitive shipboard electronic systems, there is one that predominates by offering wide open access. Not only is it the least resistant route, it is made intentionally so because its very purpose is to intercept and efficiently to collect electromagnetic energy from the environment. That path, of course, is through the many shipboard antenna systems, and especially through those antennas designed to operate below 100 MHz. Since the highest EMP energy products immediately appear at the antenna terminations, the first system components that require protection are the base insulators, the matching and tuning networks, and the coaxial transmission lines.

EMP energy gaining entry by way of shipboard antennas will, if the transmission lines are left unguarded, impose a sudden transient of excessive level at equipment input stages. To prevent this potentially catastrophic occurrence, techniques must be devised instantaneously to provide an alternative path for surge current flow, normally in the form of an immediate shunt to ground, wherever the applied level at the input terminals exceeds a specified threshold. But the moment the overvoltage ceases to exist, normal system operation must resume automatically. Moreover, the circuit protective device should not in any way adversely affect the performance of the system being protected.

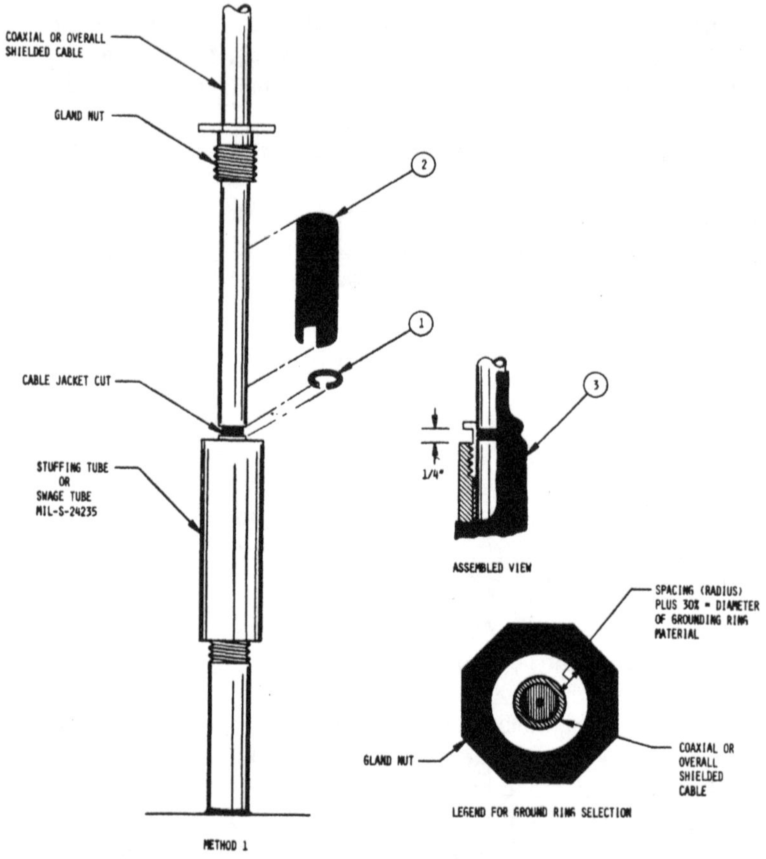

LIST OF MATERIAL			
ITEM NO.	PART	SPECIFICATION	NOTE
1	GROUNDING RING	COMMERCIAL	4, 5, 8
2	COMPRESSION SLEEVE, SHIM STOCK	COMMERCIAL	7

NOTES:

1. THIS METHOD OF CABLE SHIELD GROUNDING APPLIES TO NEW CABLE INSTALLATIONS AND TO EXISTING INSTALLATIONS WHERE THE CABLE CANNOT BE REMOVED. THIS METHOD OF CABLE SHIELD GROUNDING IS PREFERRED DUE TO LOWER COST AND SIMPLICITY OF INSTALLATION. GROUNDING EFFECTIVENESS IS APPROXIMATELY THE SAME.

2. UNSCREW PACKING GLAND NUT FROM THE STUFFING TUBE AND MOVE IT SEVERAL INCHES UP THE CABLE AND TAPE. THE CABLE IS NOT REQUIRED TO BE REMOVED FROM THE CABLE HANGERS OR TO BE MOVED IN ANY WAY.

3. WITH A POCKET KNIFE OR SIMILAR TOOL, MAKE TWO CIRCULAR CUTS IN THE CABLE JACKET, ONE APPROXIMATELY FLUSH WITH THE TOP OF THE STUFFING TUBE OR SWAGE TUBE AND ANOTHER APPROXIMATELY ONE-FOURTH INCH HIGHER. ENSURE INSIDE OF GLAND NUT IS CLEAN AND FREE OF PAINT, SEALING COMPOUNDS, OR CORROSION. CLEANING WITH FINE SANDPAPER MAY BE REQUIRED.

4. REMOVE THE CUT SECTION OF THE CABLE JACKET. SELECT PROPER GROUNDING RING MATERIAL IN ACCORDANCE WITH THE LEGEND. CUT GROUNDING RING MATERIAL TO PROPERLY FIT (BUTT END-TO-END) THE CABLE AREA WHERE JACKET WAS REMOVED. COAT THE GROUNDING RING WITH ANTI-SEIZE COMPOUND OF MIL-T-22361. APPLY A COATING OF ANTI-SEIZE COMPOUND TO THE EXPOSED CABLE SHIELD AND TO THE THREADS OF THE GLAND NUT.

5. PLACE GROUNDING RING AROUND CABLE IN AREA WHERE JACKET WAS REMOVED. PLACE COMPRESSION SLEEVE AROUND CABLE JACKET AND GROUNDING RING. HOLDING COMPRESSION SLEEVE TIGHTLY AROUND CABLE AND GROUNDING RING, SLIDE GLAND NUT DOWN OVER SLEEVE AND THREAD INTO STUFFING TUBE. AFTER THREADS HAVE ENGAGED AT LEAST ONE-FOURTH INCH, REMOVE THE COMPRESSION SLEEVE AND COMPLETE TIGHTENING OF THE GLAND NUT AS REQUIRED FOR PACKING. WHEN COMPLETED, THE GROUNDING RING SHOULD BE LOCATED APPROXIMATELY MIDWAY ON THE INSIDE OF THE GLAND NUT WITH AN APPROXIMATE ONE-FOURTH INCH GAP BETWEEN THE TOP OF THE STUFFING TUBE AND THE BOTTOM OF THE HEX ON THE GLAND NUT. ADDITIONAL PACKING OR GLAND WASHERS MAY BE REQUIRED.

6. WEATHERSEAL AS SPECIFIED FOR CORROSION PROTECTION.

7. THE COMPRESSION SLEEVE IS USED ONLY TO COMPRESS THE GROUNDING RING WHILE REINSTALLING THE GLAND NUT. IT CAN BE CUT FROM APPROXIMATELY 0.005" MATERIAL.

8. ENSURE ELECTRICAL CONTACT BETWEEN GROUNDING RING AND CABLE SHIELD ON CORRUGATED OUTER SHIELD CABLES.

9. WHERE METAL PIPES AND TUBING ARE ROUTED THROUGH STUFFING TUBES, THE METHOD SHOWN HERE FOR CABLE SHIELD GROUNDING SHALL ALSO BE USED FOR PIPES AND TUBING GROUNDING.

10. GROUNDING RING SHALL BE ROUND CROSS SECTION, NEOPRENE SPONGE, FERREX WIRE, EMI MESH STRIP, CHROMERICS, INC., CARSON, CA, OR EQUAL.

Figure 6-13(a) Cable Shield Grounding Methods

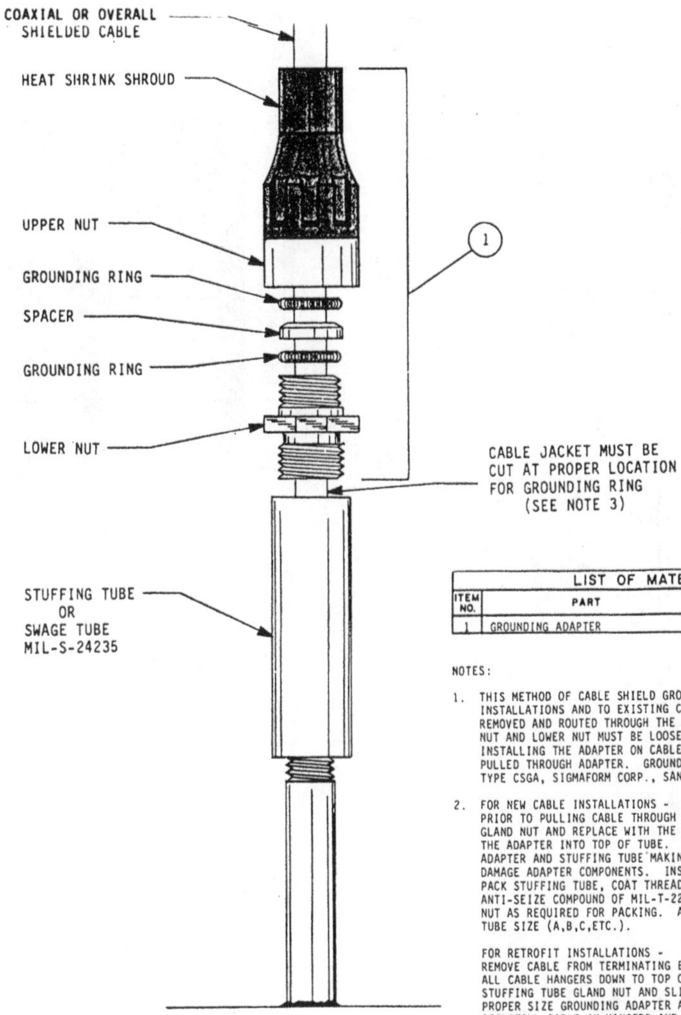

METHOD 3

LIST OF MATERIAL			
ITEM NO.	PART	SPECIFICATION	NOTE
1	GROUNDING ADAPTER	COMMERCIAL	1

NOTES:

1. THIS METHOD OF CABLE SHIELD GROUNDING APPLIES TO NEW INSTALLATIONS AND TO EXISTING CABLES THAT CAN BE REMOVED AND ROUTED THROUGH THE GROUNDING ADAPTER. UPPER NUT AND LOWER NUT MUST BE LOOSELY THREADED TOGETHER WHEN INSTALLING THE ADAPTER ON CABLE TO ALLOW CABLE TO BE PULLED THROUGH ADAPTER. GROUNDING ADAPTERS SHALL BE TYPE CSGA, SIGMAFORM CORP., SANTA CLARA, CA, OR EQUAL.

2. FOR NEW CABLE INSTALLATIONS -
PRIOR TO PULLING CABLE THROUGH THE STUFFING TUBE, REMOVE GLAND NUT AND REPLACE WITH THE ADAPTER, LOOSELY THREADING THE ADAPTER INTO TOP OF TUBE. PULL CABLE THROUGH ADAPTER AND STUFFING TUBE MAKING SURE CABLE DOES NOT DAMAGE ADAPTER COMPONENTS. INSTALL CABLE IN HANGERS. PACK STUFFING TUBE, COAT THREADS OF LOWER NUT WITH ANTI-SEIZE COMPOUND OF MIL-T-22361, AND TIGHTEN LOWER NUT AS REQUIRED FOR PACKING. ADAPTER CHOSEN MUST MATCH TUBE SIZE (A,B,C,ETC.).

   FOR RETROFIT INSTALLATIONS -
REMOVE CABLE FROM TERMINATING EQUIPMENT AND REMOVE FROM ALL CABLE HANGERS DOWN TO TOP OF STUFFING TUBE. REMOVE STUFFING TUBE GLAND NUT AND SLIDE OFF OF CABLE. CHOOSE PROPER SIZE GROUNDING ADAPTER AND SLIDE DOWN OVER CABLE REINSTALL CABLE IN HANGERS AND RECONNECT TO TERMINATING EQUIPMENT. COAT THREADS OF LOWER NUT WITH ANTI-SEIZE COMPOUND OF MIL-T-22361. THREAD ADAPTER INTO STUFFING TUBE AND TIGHTEN LOWER NUT AS REQUIRED FOR PACKING.

3. AFTER CABLE HAS BEEN PERMANENTLY INSTALLED IN PLACE AND ALL HANGERS ARE TIGHTENED, UNSCREW UPPER NUT AND MOVE IT SEVERAL INCHES UP THE CABLE AND TAPE. MAKE TWO CIRCULAR CUTS IN CABLE JACKET, ONE FLUSH WITH THE TOP OF LOWER NUT AND ANOTHER ONE-HALF INCH UP.

4. REMOVE CUT SECTION OF CABLE JACKET AND APPLY A COATING OF ANTI-SEIZE COMPOUND OF MIL-T-22361 TO THE EXPOSED CABLE SHIELD, TO THE GROUNDING RINGS, AND TO ALL ADAPTER THREADS. LOWER THE UPPER NUT AND HAND-TIGHTEN, MAKING SURE THE GROUNDING RINGS FALL INTO THE SLOT CUT IN THE CABLE JACKET.

5. WEATHER SEAL BY APPLYING HEAT TO TOP OF WEATHER SHROUD AS REQUIRED FOR PROPER SHRINKAGE AROUND CABLE. AFTER INSTALLATION, PERIODIC TIGHTENING OF THE ADAPTER FOR CABLE WEATHER SEALING IS NOT REQUIRED.

Figure 6-13(b) Cable Shield Grounding Methods

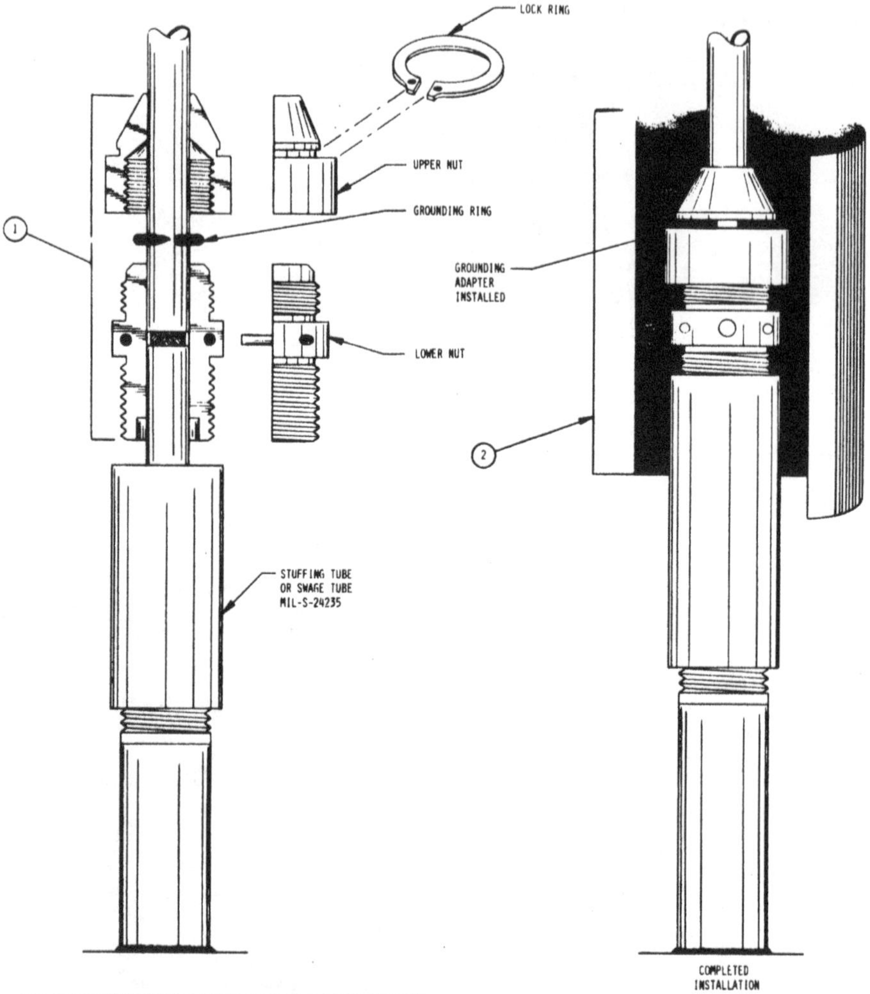

LIST OF MATERIAL			
ITEM NO.	PART	SPECIFICATION	NOTE
1	GROUNDING ADAPTER	COMMERCIAL	1
2	SLEEVE, CABLE REPAIR	MIL-I-23053/15	8

NOTES:

1. THIS TYPE GROUNDING ADAPTER IS FOR RETROFIT USE ONLY WHERE THE CABLE CANNOT BE REMOVED TO INSTALL THE UNSPLIT TYPE GROUNDING ADAPTER. THIS SPLIT GROUNDING ADAPTER SHALL BE TYPE CSGA(S), SIGMAFORM CORP., SANTA CLARA, CA, OR EQUAL.

2. UNSCREW PACKING GLAND NUT FROM STUFFING TUBE AND CUT WITH HACKSAW OR OTHER TOOL, REMOVE FROM CABLE AND DISCARD.

3. SELECT GROUNDING ADAPTER TO FIT STUFFING TUBE AND DISASSEMBLE FOR INSTALLATION.

4. REASSEMBLE LOWER NUT AROUND CABLE, COAT ALL THREADS WITH AN ANTISEIZE COMPOUND OF MIL-T-22361, AND THREAD NUT INTO STUFFING TUBE. TIGHTEN AS REQUIRED FOR PACKING.

5. USING A POCKET KNIFE OR SIMILAR TOOL, CUT THE CABLE JACKET AROUND THE PERIMETER OF THE CABLE FLUSH WITH THE TOP OF THE LOWER NUT. MAKE ANOTHER CUT IN THE CABLE JACKET APPROXIMATELY ONE-FOURTH INCH HIGHER AND REMOVE THE CUT SECTION OF CABLE JACKET.

6. COAT THE EXPOSED CABLE BRAID WITH THE ANTISEIZE COMPOUND IDENTIFIED IN NOTE 4.

7. ASSEMBLE GROUNDING RING AND UPPER NUT AROUND CABLE. LOCK UPPER NUT SECTIONS TOGETHER WITH SNAP RING. PLACE THE GROUNDING RING AROUND THE CABLE WHERE JACKET WAS REMOVED AND HAND TIGHTEN UPPER NUT TO LOWER NUT.

   -DO NOT OVER-TIGHTEN-

8. PLACE CABLE REPAIR SLEEVE AROUND COMPLETE ADAPTER AND HEAT SHRINK AS REQUIRED FOR WEATHER SEALING, ENSURING AREA AROUND CABLE IS THOROUGHLY SEALED. PERIODIC TIGHTENING OF THE ADAPTER FOR CABLE WEATHER SEALING IS NOT REQUIRED. INSTALL CAUTION TAG

Figure 6-13(c) Cable Shield Grounding Methods

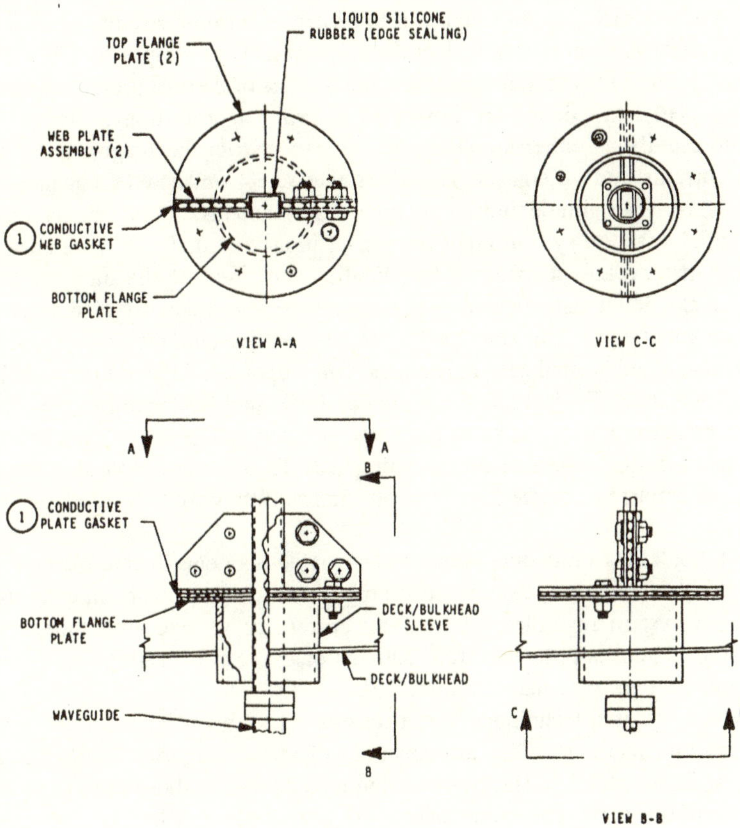

LIST OF MATERIAL			
ITEM NO.	PART	SPECIFICATION	NOTE
1	GASKET MATERIAL, CONDUCTIVE	COMMERCIAL	1

NOTES:

1. CONDUCTIVE GASKET MATERIAL SHALL BE CHROMERICS, INC., CARSON, CA, "POLASHEET", OR EQUAL. THE MOUNTING SURFACES FOR THE GASKET SHALL BE CLEANED TO BRIGHT METAL AND COATED WITH ANTI-SEIZE COMPOUND OF MIL-T-22361 PRIOR TO INSTALLING GASKET.

2. SPLIT SLEEVE INSTALLATION SHOWN IS AS DETAILED IN EIMB SERIES, NAVSEA 0967-000-0110.

Figure 6-14 Waveguide Grounding

Various mechanisms employed to protect electronic circuits against EMP overload are known as terminal protection devices, or simply as TPDs (some sources define TPDs as transient protection devices or thermal protection devices; the accepted Navy definition, however, is terminal protection devices). TPDs operate mainly as either amplitude limiters to restrict the magnitude of high-level currents and voltages, or as filters that reject undesired frequency components. Most amplitude limiters are high-impedance insulators, usually installed in parallel with the system input lines or at input terminals. When a current or voltage that exceeds a specified threshold is impressed on the line, the limiter TPD breaks down instantaneously to offer a very low resistance path away from the equipment input. This shunt path remains operative until the excessive voltage is dissipated and normal current resumes, whereupon the TPD returns to its high-impedance state. Typical amplitude limiter TPDs include metal-oxide varistors, semiconductor diodes, and spark gap surge arresters, all of which must be rugged enough in design to withstand the extreme EMP intensity. Note that varistors and diodes must be designed for low capacitance if they are to be used in antenna circuits.

Filter TPDs strain out specific frequency components in an undesired high-energy pulse spectrum, thereby preventing those frequencies from entering the protected system regardless of transient current or voltage amplitude. In this case, the system is protected from damage or upset even from pulse transients not high enough to actuate a limiter TPD.

Other related techniques useful for dealing with EMP are known as transient-tolerant methods. They are generally of three categories: hardware, software, and procedural.[10] Hardware design techniques are those used to preclude both transient damage to components and system upset, whereas software and procedural methods protect only against system upset. Hardware devices include switches, transformers, relays, chokes, circuit breakers, and redundant system elements. Software techniques incorporate error detection and correction codes with built-in data transmission error toleration to circuit upset. Procedural methods involve training and operational skills to recognize and recover quickly from the effects of EMP.

Because of the extremely rapid rise time of EMP transients, electromechanical devices such as switches, relays, and circuit breakers are simply unable to respond quickly enough to ward off trouble. Therefore, spark gaps, gas-discharge arresters, and semiconductor diodes are favored for naval shipboard EMP protection. However, semiconductors are nonlinear, and, as such, are potential generators of intermodulation interference, so they are used only in locations other than antenna terminals where signals are low enough not to promote the generation of intermodulation products. Also, spark gaps used ordinarily for lightning protection in shipboard antennas are too slow to react to EMP transients. Standard lightning arresters normally fire at a few hundred volts with a typical lightning stroke rise time of microseconds but the rise time for EMP energy is about 10 ns with an amplitude of several kilovolts. Thus, lightning arresters provide very poor protection against EMP.[11] As a result, the preferred shipboard TPDs are specialized fast-operation spark gaps pressurized with a trace of low-level radiation gas. Such TPDs are known as gas tubes or gas-discharge surge arresters. Since they are clean of intermodulation products, until fired, they are the most suitable TPD choice for protection of shipboard antenna terminals.

The gap breakdown and discharge characteristics of a gas arrester depend on the type of gas (typically low-pressure argon or hydrogen), gas pressure, gas temperature, shape of gap points, gap length (usually two electrodes spaced a few tenths of a millimeter apart inside a ceramic tube), and the nature of the applied voltage. Specifications for gas TPDs are in terms of dc breakdown voltage and maximum current handling capabilities. Transient impulse breakdown voltages are higher than the dc breakdown limit, but, of course, must be below the maximum peak safe level of the circuit being protected.

In most cases, it is impractical to install the TPD at the antenna feed point. It is more convenient to place it at the high voltage insulator (or, if possible, to mount it inside the insulator). Again, the TPD must not be allowed to degrade the normal performance of the protected system; for example, by increasing the VSWR. It must bear well the normal rigors of the shipboard environment, such as shock, vibration, temperature, and humidity. Finally, TPD insertion loss should not exceed 1 dB, and TPD intermodulation products must be 80 dB down.

6-4 EMP Testing and Modeling

With test ban treaties presently existing among the major world powers, and with heightened public concern over nuclear safety, it is quite unlikely that there will be any detonation of nuclear weapons in the earth's atmosphere, short of war conditions. Consequently, the effects of actual nuclear-generated EMP cannot be tested. Several government, military, and private industry facilities currently exist to conduct EMP simulation studies. To observe the potential effects of EMP on shipboard electronics, the Navy owns a full-scale test range called EMPRESS and an EMP protection design and assessment program having the acronym EMPAL.

6-4.1 EMPRESS Testing

As a means to evaluate the behavior and survivability of ship systems, as well as to identify vulnerable electronic circuits and to develop the necessary technology for hardening ships against the effects of EMP, the Navy operates the Electromagnetic Pulse Radiation Environment Simulator for Ships. Known familiarly as EMPRESS, the test range is near the Chesapeake Bay at Solomons, Maryland.[12] At present it is the only EMP simulation range in the world for ships. With no opportunity likely in the foreseeable future for high-altitude nuclear explosions, whole-ship testing in an EMP simulator is necessary to assess potential effects on the combat readiness of a ship. By using EMPRESS, full-system tests are carried out to determine where, and how much, EMP energy is ultimately conducted to critical points inside a ship, and whether the ship can continue to operate effectively.

The ability of EMP to transfer energy to a system is a function of the pulse spectrum excitation frequencies. That is, EMP has the potential to affect any electrical or electronic system which operates anywhere within the very wide EMP spectrum. EMPRESS is specially designed to imitate as closely as possible the EMP spectrum projected in Figure 6-15. The EMPRESS test range simulates nuclear-generated EMP by energizing high-voltage pulse emissions. As pictured in Figure 6-16, a long-wire antenna is used to radiate horizontally polarized pulses, and an inverted cone transmits vertically polarized pulses. The horizontally polarized energy produces RF currents along horizontal structures of the ship hull, while the vertically polarized energy couples to vertical members such as masts, stacks, weapons, and superstructure. Measurements collected during the tests are then extrapolated upwards to predict the real effect of an actual nuclear EMP intrusion.

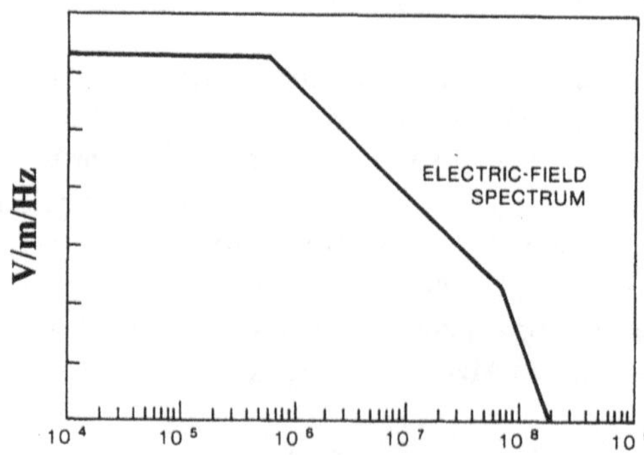

Figure 6-15 Frequency Spectrum of High-Altitude EMP

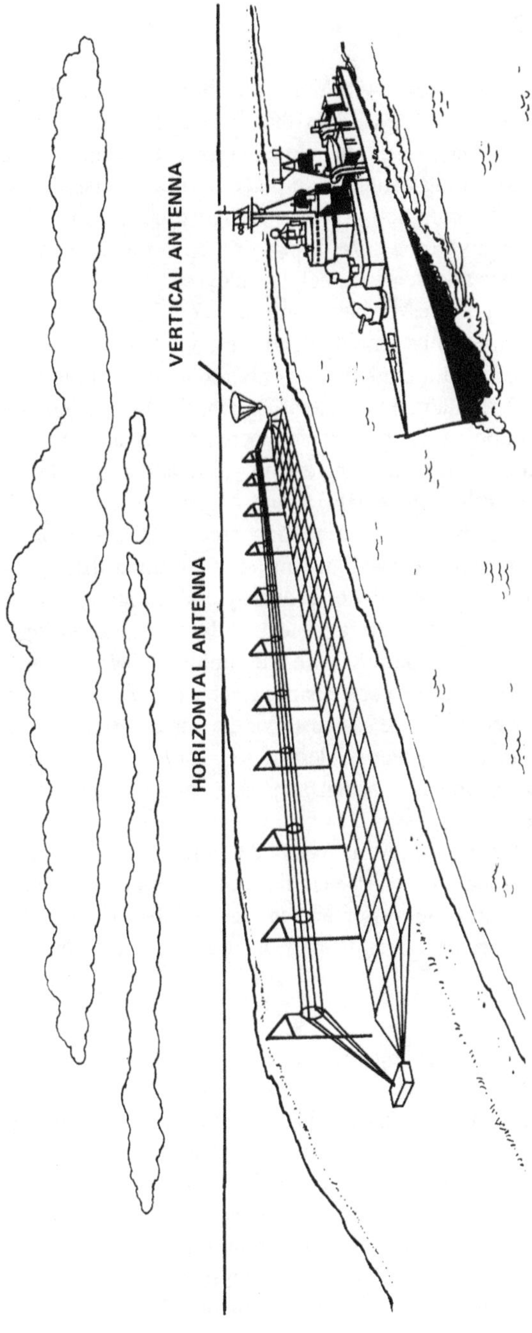

Figure 6-16 EMPRESS Antennas

6-4.2 EMP Modeling

There are problems with full-scale ship illumination by an electromagnetic simulator such as EMPRESS in that it can be quite costly, both in time and money, and it may result in some damage to vital ship electronic circuits. Moreover, the testing cannot take place until after the ship has been built; that is, long after the completion of the design process. In recognition of the need for EMP protection to be done as part of the overall ship design, the Navy makes use of computer and scale modeling analyses.

Assuming that EMP hardening of a ship has been satisfactorily dealt with through such means as shielding, grounding, bonding, and filtering at all hull aperture and penetration points, the only remaining opportunity for EMP invasion is through the ship antenna systems. Accordingly, modeling efforts are devoted to the protection of equipment connected to the antenna. The principal Navy modeling program for assessing circuit vulnerability and development of antenna protection devices is the EMP Design Algorithm, or EMPAL.[13]

Prior to activating the modeling process, there are several considerations to take into account: First, if it can be determined that no protective device is needed, then, in the interest of economy and design simplicity, none should be used. Second, if the need for a protective device is established, then the proposed solution must be compatible with the ambient EME of the ship and must not adversely affect system performance. For example, it would be unfortunate to have a TPD needlessly fire as a result of the normal RF conditions in the shipboard EME. Finally, the protective design solution should be accomplished early in the ship design process so that ship acquisition is not in any way delayed by EMP suppression plans.

To perform a circuit analysis, the anticipated EMP voltages and currents and their time behavior characteristics must be modeled. The susceptibility analysis then becomes the systematic process of determining the relative hardness of electronic circuits to EMP-induced transients and the probable damage or upset.

6-4.2.1 EMPAL Analysis Process

A shipboard antenna system is described by an antenna element in a specific topside location and an associated RF system that connects to the antenna. The antenna element is viewed as a transfer function represented by an equivalent circuit. The RF system elements are represented by equivalent circuits comprising conventional circuit elements. For the purpose of this analysis, the EMP environment drives the circuit. The EMP is represented by a mathematical expression such as the sum of exponential functions that decay with time. Along with this information, a circuit simulation computer program is used to determine the time or frequency response at individual nodes in the circuit.

A ship system procedure that meets the TPD design approach requirements is shown as Figure 6-17. An essential analysis tool of the EMPAL system is the circuit simulation computer program which determines circuit response by solving an electrical circuit represented in mathematical terms. Time-domain response determination (transient analysis) is a function of circuit analysis. A system of equations to describe a circuit is determined by the nodal equations for each element and the constraints of element interconnection. The constraints reflect Kirchoff's current and voltage law, and the circuit equations consist, in general, of a system of differential equations. The requirement of circuit analysis is to determine the solution of this system of equations for various conditions.

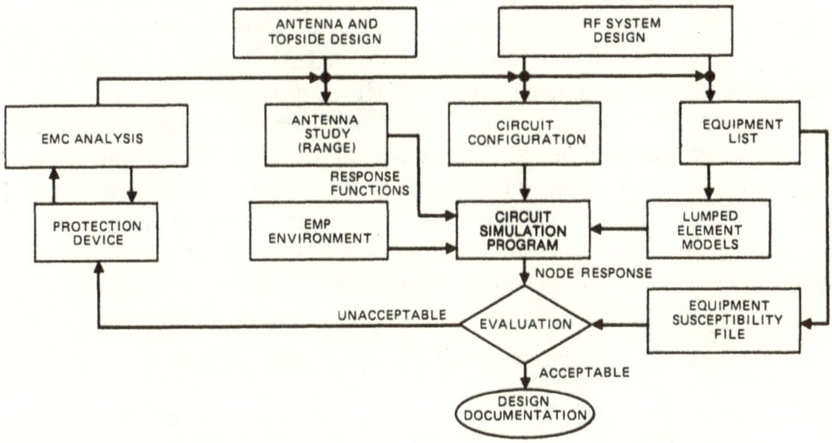

Figure 6-17 EMPAL Design Process

The nodal responses are compared against equipment susceptibilities. An evaluation is then made of the need for increased EMP protection. If the nodal responses exceed the susceptibility levels of the system hardware components, the system is considered to be vulnerable and protection is placed at the antenna terminals. Any TPD to be employed must be evaluated through EMC analysis before its installation to ensure that it will not be fired by the ambient environment of the ship and that it will not seriously degrade RF system performance. A later analysis is made with the TPD included in the circuit configuration. If the nodal responses are below the susceptibility levels of the system hardware components, the design is acceptable. As a final step of the EMPAL design process, the design is documented to include a functional specification of any required TPD.

6-4.2.2 Scale Modeling Process

Shipboard topsides are so complex that EMPAL numerical modeling alone is seldom sufficient. Consequently, a secondary method of scale-modeled measurements has proven highly successful for design and development of EMP protection requirements. Scale brass ship models are routinely constructed as part of the HF antenna arrangement design effort associated with new ship construction and existing ship alteration programs. Currently, these models support frequency-domain measurements for determining antenna design and performance characteristics. The same models are suitable for time-domain measurements because they are faithfully constructed to very fine detail. Therefore, they permit accurate data to be obtained over the entire frequency range of interest for EMP. By comparing scale-model with full-scale measurements, it has been shown that the scale-model approach is a cost-effective means of providing information concerning the coupling of EMP to complex metallic structures. Consequently, the Navy has constructed a scale-model transient range that can be used in conjunction with the EMPAL design process.

The bounded-wave simulator pictured in Figure 6-18 creates an imitation of a vertically polarized plane wave incident on the scale model ships at zero-degree elevation. Almost all shipboard HF antennas are vertically polarized and have low-elevation main lobe patterns at frequencies corresponding to high energy levels in the EMP spectrum. Hence measured data from the bounded-wave simulator are suitable for worst-case vulnerability analysis.

Figure 6-18(a) Bounded–Wave EMP Modeling Simulator (Overall View)

Figure 6-18(b) Bounded–Wave EMP Modeling Simulator (Close-up Expanded View)

REFERENCES

1. Jack Anderson, "Awesome EMP Could End War Before It Starts," *The Washington Post*, 3 December 1982.
2. "U.S. Seeks Answers to A-Blast Oddity," Electronic News, October 30, 1967.
3. R. V. Carstensen, "An Approach to EMP Hardening for Naval Ships," *Naval Engineers Journal*, April 1979, p. 144.
4. EMP Awareness Course, Defense Nuclear Agency, Topical Report DNA 2772T, Washington, DC, October 1977, pp. 3-1 through 3-10.
5. L. W. Pinkston, "Electromagnetic Pulse (EMP)," *Interference Technology Engineers Master*, Robar Industries, West Conshohocken, PA, 1984, p. 36.
6. M. A. Rose and L. Duncan, "EMP Susceptibility Analysis," *Interference Technology Engineers Master*, Robar Industries, West Conshohocken, PA, 1984, p. 40.

7. *Ibid.*, p. 48.
8. R. V. Carstensen, *Electromagnetic Interference Control in Boats and Ships*, Don White Consultants, Gainesville, VA, 1981, p. 6.7.
9. D. S. Dixon and S. I. Sherman, ''Reduction of EMI/EMP Shield Currents Using an Improved Shield Ground Adapter,'' IEEE International Symposium on Electromagnetic Compatibility Symposium Record 1986, p. 428.
10. Pinkston, *op. cit.*, p. 44.
11. K. G. Cook, ''Gas-filled Surge Arresters for NEMP Protection,'' *Interference Technology Engineers Master*, Robar Industries, West Conshohocken, PA, 1986, p. 85.

Chapter 7
Shipboard Electromagnetic Assessment (EMA)

7-0 THE NEED FOR PREDICTIVE ANALYSIS

Having reached this point in the book, the reader must by now have a keen appreciation of the design complexities and engineering processes required to attain EMC, to control EMI, and to provide for optimum electromagnetic performance of shipboard systems. We have reviewed the origination and long history of interference to naval electronics in Chapter 1. We have, in Chapter 2, described the involved shipboard EME. In Chapter 3, we have examined the concept of EMC and the need for effective engineering management to provide for EMC in ship design. We have explored in detail the many forms of shipboard EMI and ways to eliminate it, or at least to reduce it to a minimum, in Chapter 4. In Chapters 5 and 6, we have determined the best methods to deal with the problems of shipboard EMR hazards and the threat of EMP. Still, there remains one other important part to the whole: Can we, having intimate knowledge of the shipboard environment, armed with all the technical characteristics of the many systems to be installed, and diligently applying the principles of EMC engineering, predict with any degree of certainty that we indeed can achieve adequate shipboard EMC? How do we assess the expected degree of EMC integrity and the performance merit of electromagnetic systems in support of ship missions?

Without question such an assessment is of critical necessity, for, failing to achieve EMC during the design process, corrective action undoubtedly will be required during fleet operations later at great cost in time, engineering effort, and money. Consequently, it is far better to employ prediction and assessment techniques during the planning, design, development, and integration of shipboard electrical and electronic systems. These techniques permit identification of and economical solutions to electromagnetic problem areas before the ship goes to sea, rather than after. In other words, if potential performance degradation, electromagnetic incompatibilities, and interference conditions can be successfully predicted soon enough, then design remedies can be applied efficiently.

The alternatives to predictive analysis are not attractive. The most common alternative is the corrective action or problem solution approach at times favored by managers in the interest of meeting delivery and budget schedules. In this approach, EMC concerns are often brushed aside as the ship design is pushed toward construction, with the notion that any EMI problems that may crop up later can be resolved on a case-by-case backfit basis. This is the so-called reactive mode; i.e., patching up problems as they arise.

The other alternative is the rigid specifications approach, a method that tends to impose strict adherence to predetermined emission and susceptibility levels. Meeting these levels almost certainly will result in obtaining a more compatible system performance than the corrective action backfit approach; it might very well, however, also result in expensive overdesign by the application of solutions where problems do not exist, merely to satisfy the specification requirements.[1]

Therefore, the predictive analysis process is the option best suited to ensure that all aspects of EMC are considered while carrying out the engineering design and integration to meet mission requirements. In this manner, performance prediction is involved from the start of design and progresses in concert to installation and test.

7-1 PREDICTIVE ANALYSIS TECHNIQUES

Prediction and analysis begin with a compiling of pertinent data and study of: (1) characteristics of the operational electromagnetic environment, (2) mission requirements, (3) technical parameters of the equipment and systems to be installed, (4) emitter and sensor siting requirements, (5) emission power levels, and (6) receiver susceptibility characteristics.

The primary aim in the predictive process is to determine the electromagnetic interaction among the many electromagnetic systems and equipment of the shipboard platform; i.e., to make an *intrasystem* analysis. However, the process must also include potential EMI problems between systems of the ship platform and the various elements of other systems or platforms likely to be operating in the same general area: that is, the *intersystem* interaction. Moreover, the analysis must take into account the electromagnetic interaction between elements of the shipboard systems and the operational environment. Initiating this process early in the design affords good opportunity to influence the assignment of operating frequencies, the allocation of transmitter power levels, the placement of antennas, the arrangement of weapon systems, and, of critical importance, the identification of potential performance deficiencies.

Prediction and analysis must rely on the known or assumed electromagnetic characteristics of each individual shipboard system element and the manner in which elements may be expected to interact with each other. There are many probable self-interference coupling paths available in the ship, including antenna-to-antenna, cable-to-cable, cable-to-equipment, and equipment enclosure-to-equipment enclosure. The chief mode of interaction aboard ship, however, is coupling of radiating energy from transmitting antennas to receiving antennas. The situation is exacerbated by the large number of systems typically required to operate simultaneously in the congested ship. When the entire electronic suite of a modern warship is energized, the topside becomes a time-varying electromagnetic entity, and system performance becomes extremely sensitive to the nature and spatial relationship of each element with respect to every other and to the overall topside arrangement. The electromagnetic assessment, therefore, must systematically account for the mutual interaction among electronic equipment, ship structures, operation dynamics, and the environment to determine the interactive effects.[2]

To assemble the necessary data for prediction and analysis, the following factors should be determined:

a. Each system or unit of equipment which may influence EMC, whether the item is active or passive, and potential problem areas that are either inherent or definable.
b. The historical record of EMI problems experienced in the fleet in similar configurations, and the corrective remedy applied.
c. The various operating frequencies and probable effects among the equipment and systems.
d. Which of several locations for emitters and sensors should provide the least interference.
e. All potential sources and causes of known EMI problems and whether they are time-varying or steady-state contributors.
f. The type and degree of suppression required for corrections.
g. Susceptibility characteristics for each sensor, including minimum threshold response in amplitude and duration.
h. The purpose of each system and whether it is to be operated continuously or intermittently.
i. The criticality of each system to overall ship mission requirements.
j. The effect of the shipboard structural environment and equipment or systems other than electromagnetic.

The process of prediction and analysis naturally is complicated by many uncontrolled and often unexpected interference factors that enter situations, such as antenna sidelobe, backlobe, and reflected radiation; spurious and harmonic signal leakage; and production of intermodulation noise.

Through a combination of experience and expertise it is possible in some cases to predict fairly accurately the degree of performance and EMC achievable. The extreme complexity of naval shipboard electronic systems, however, and the need to compile great amounts of frequency spectrum and technical parameters for the numerous units of equipment make the use of computer modeling the only truly practical means to obtain realistic assessments of system performance and EMC integrity.[3]

7-2 ELECTROMAGNETIC ASSESSMENT MODELING

To model the interaction among shipboard electromagnetic systems, the three essentials are: (1) an electromagnetic energy source, (2) an electromagnetic coupling mechanism, and (3) an electromagnetic receptor. In general, the primary sources of shipboard electromagnetic energy are the several onboard emitters radiating intentionally. Secondary sources are reradiation of coupled energy from cables and scattering from metal structures, and noise generation of intermodulation, spurious, and harmonic signal products. Electromagnetic coupling methods are principally direct electromagnetic propagation paths, and a variety of reradiation paths of RF coupling energy from cables, waveguides, and metal objects. Receptors, in virtually all instances, are the ship's receiving equipment. There are other devices, however, that act as unintentional susceptors; e.g., the pressure- and level-indicating transducer sensors used in electronic automatic control circuits.

To model the electromagnetic energy sources, emitter frequencies and power levels are needed. For receptors, the needed technical characteristics are receiver sensitivity, selectivity, and response time. The normal modeling procedure is to select a possible receptor and a probable EMI source, then examine the various amounts of energy received over the many potential coupling mechanisms. The process is repeated for all possible sources, and the resultant performance degradation is determined. A new receptor is selected and the routine iterated until all potential source-victim pairs have been explored. From the resulting findings, an EMI matrix is drawn up as typified by Table 7-1. Solutions are proposed to avoid or to minimize the incompatibilities so that satisfactory EMC is reached. As a consequence, an orderly, systematic prediction of potential EMI problems and recommended solutions are implemented as an integral part of the system design process from its commencement. Three of the most-used modeling programs in US naval shipboard electromagnetic assessment today are Shipboard EMC Analysis-Communications (SEMCAC), Shipboard EMC Analysis-Microwave (SEMCAM), and Topside Design Model (TDM). Functional details and operating procedures for these programs literally fill volumes; therefore, only an overview of program capabilities will be given here.

Table 7-1. Potential EMI Problems and Recommendations

VICTIM (RECEPTOR) / *Source (Emitter)*	HF Communication	VHF Communication	UHF Communication	SATCOM Receive	SATCOM Transceive	IFF	Glide Slope Indicator	NAVSTAR GPS Navigation	OMEGA Navigation	Surface Search Radar	Navigation Radar	Weapons Control Radar	Electronic Warfare Suite
HF Communication	1A 6G	2A 6G	2A 6G	2A 3A,F			5J			3A,G *		5F	5F,G 1
VHF Communication		1A	2A										
UHF Communication			1A	1A	1A								
SATCOM Transceive			1A,N										
IFF						1P							
Surface Search Radar													1B
Navigation Radar													1B
Weapons Control Radar													1B,D

EMI Category

1. Equipments operate in same frequency band
2. Harmonic relationship
3. Transmitter can operate at receiver IF
4. Adjacent frequency-equipment responds to high power or spurious outputs
5. Responds to out-of-band frequencies (case/cable penetration)
6. Intermodulation interference (IMI)
7. Unfocused noise (EMC only)
8. Broad Band Noise (BBN)
9. Superstructure reflection

EMI Fix

A. Frequency Management
B. Blanking
C. ECP 325 (Band 1 Blanking) (EW Suite)
D. ECP 327 (High rep rate blanking) (EW Suite)
E. Problem and fix under investigation
F. Approved EMI filters or additional shielding
G. Bond/ground MIL-STD-1310
H. Frequency sectoring (software program) (EW Suite)
I. ECP 312 (metallized enclosure) (EW Suite)
J. NAEC FC 19 & 20 (Glide Slope Indicator)
K. ECP 322 (CW EMI) (EW Suite)
L. Proper antenna isolation
M. Install RAM/relocate antenna
N. Install/use antenna multicouplers
O. ORDALT 15204
P. Defruiter used to eliminate Transponder to Interrogator EMI
Q. Repair/Realign Equipment

\+ Reported in 1 instance
* Fix probably not required

7-2.1 SEMCAC Modeling

The SEMCAC computer program is used extensively in ship design projects to analyze the EMC of shipboard communications systems and antennas. From the analyses, communications circuit performance is predicted. SEMCAC retrieves data base information, groups functional models of the communications systems, culls signal frequencies and amplitudes, and duplicates detector performance. The program output results in design recommendations, communications performance ranges, and an EMC management plan.

A subsidiary of SEMCAC is the Antenna Scattering Analysis Program (ASAP), which predicts communications antenna radiation patterns and impedance in the shipboard EME, recommends optimum siting of communications antennas for EMC, determines communications frequency restrictions, and plots radiation hazard contours for communications transmitting antennas.

7-2.2 SEMCAM Modeling

The SEMCAM program is a computer model for evaluation of antenna-to-antenna coupled interference among shipboard microwave systems. SEMCAM predicts and evaluates the effects of EMI on a pair-by-pair source-victim basis. Its primary application is in the shipboard design process to assess the relative EMC merit of various microwave systems and antenna arrangements. SEMCAM provides the following outputs:

a. An identification of requirements for compatible operation.
b. An automatic cull of noninterfering pairs and a determination of qualifications for degraded performance conditions.
c. An assessment of problem severity and recommended solutions.
d. A determination of best and worst case frequency separation between source-victim pairs.
e. A determination of best and worst case equipment models.
f. A recommendation of best antenna EMC orientation between each source-victim pair.

SEMCAM programming information is derived from either measured or modeled representations of transmitter emission spectra, including spurious and harmonic emissions and empirical coupling formulas. Also, measured or modeled receiver characteristics of RF, IF, selectivity, bandwidth, noise figure, and signal processing are used.

7-2.3 TDM Performance Assessments

The Navy's TDM is a computer-aided systems engineering program currently used in all major ship designs, modernizations, and overhauls affecting shipboard electromagnetic performance. The unique advantage of TDM is that it allows immediate stage-by-stage performance assessment for each arrangement option and each change in system characteristic or locations. It is therefore used to synthesize topside arrangements of electromagnetic systems optimally and to evaluate the resultant performance.

Since the topside of a naval ship provides very limited space to place the many items necessary to support mission requirements and operational capabilities, performance compromises result as a consequence of spatial and electromagnetic interaction among the topside items. The purpose of TDM is to allow ship designers to exploit the available topside space and to extract the best performance possible from topside systems. TDM is particularly useful in the feasibility, concept, and early preliminary design phases when there is a need for examining numerous alternative candidate ship topside arrangements.

Because pictorial information has been found to be more helpful to the designer than alphanumeric printed outputs, TDM is an interactive graphics display tool. System software consists of the following major modules: (1) data library, (2) space planning, (3) physical evaluation, and (4) performance evaluation. The data bank is used to create and maintain data files with descriptions of ships, including ship hull, deckhouses, masts, and arrangeable system components. The space planning module is used to create a numerical model of the topside volume and each object arranged within this volume. It provides the basic capability to add and delete arrangeable items and supporting structures. The module may be used also to signal the designer about constraints which have been defined for the elements stored in the data file.

The physical evaluation module is used to compute topside weight distribution and center of gravity and to estimate the deck area and the volume enclosed by a deckhouse.

Lastly, the performance evaluation module is used to evaluate each candidate topside arrangement proposed. This module contains Performance Evaluation Program (PEP) subroutines, known as PEP1, PEP2, and PEP3, which have the capability to assess particular phases of the topside synthesis process. The evaluation techniques in the PEP routines are based upon principles of physics and naval ship systems information.

PEP1 provides performance estimates to aid in arranging topside elements. Its highly interactive algorithms allow the designer to assess performance during any point in the topside design process. PEP2 is used for evaluation scoring of individual topside components after an initial topside arrangement or rearrangement has been completed. PEP3 performs directive antenna gain-reduction evaluations.

As a comprehensive assessment program, therefore, TDM allows the user to evaluate electromagnetic radiated energy blockage and coverage, radar line-of-sight range detection, and communication antenna range prediction. The algorithm includes the cumulative amplitude probability distribution of HF communications and the gain reduction due principally to blockage by superstructure.

7-2.3.1 TDM Geometry

TDM requires a geometric description of the ship in the form of a three-dimensional model. This model is derived from standard Navy computer data and ship drawings. The TDM model data is made up of three-dimensional elements defining the ship's hull, deckhouse, masts, and arrangeable items. The ship is defined (i.e., ship coordinate system) by the X,Y,Z points of a predetermined Cartesian coordinate system (see Figure 7-1), where:

X = Distance in feet from the forward perpendicular (FP). X is negative for points forward of the FP and positive for all points aft of the FP.

Y = Half-beam (the positive distance in feet from the Y=0.0 ship centerline plane).

Z = Height in feet measured from the baseline.

The TDM program allows one to simulate ships grossly by use of geometrically simple forms such as rectangular prisms, circular cylinders, and hexagonal columns, and to evaluate the EM performances of the topside elements. The ship's overall EM characteristics are optimized by carefully siting and adding or deleting arrangeable topside elements on the graphic display.

First, the hull on the drawing or sketch of the candidate ship is divided into several sections and stored in the hull element file. The beamwidths and vertical lines at specific heights above the design waterline for each section are entered into the hull-building routine. After the hull-building is completed, the deckhouse is simulated with rectangular prisms by generating levels along the vertical axis and sections along the horizontal axis. Then the beamwidths are specified at each level along the traverse axis. The deckhouse elements, thus

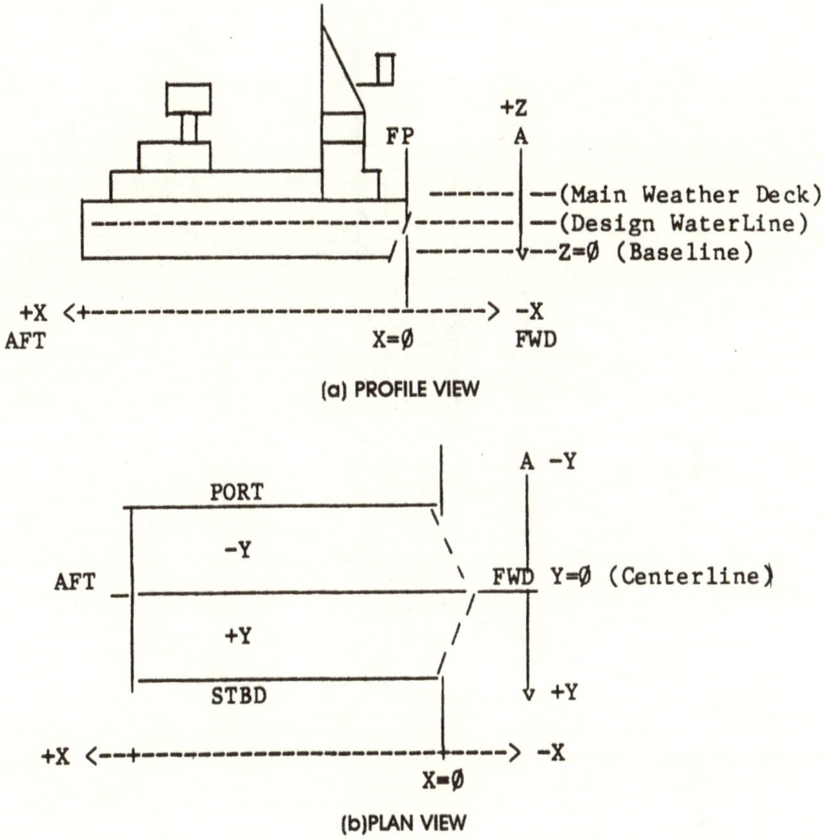

Figure 7-1 TDM Coordinate Geometry

constructed, are stored in the deckhouse file and transferred to the hull-building program, where they are placed on the hull at any desired location. The masts and yardarms are modeled by use of a number of directed line segments, representing masts, braces, yardarms, and platforms. After the hull, deckhouse, and masts are completed, the major topside elements, including weapons and electronics, are finally arranged on the superstructure, mast, and yardarm. As an example of this building-block approach to derivation of a computer-generated hull, Figure 7-2 shows an isometric TDM view of a PHM 1 Class patrol boat.[4]

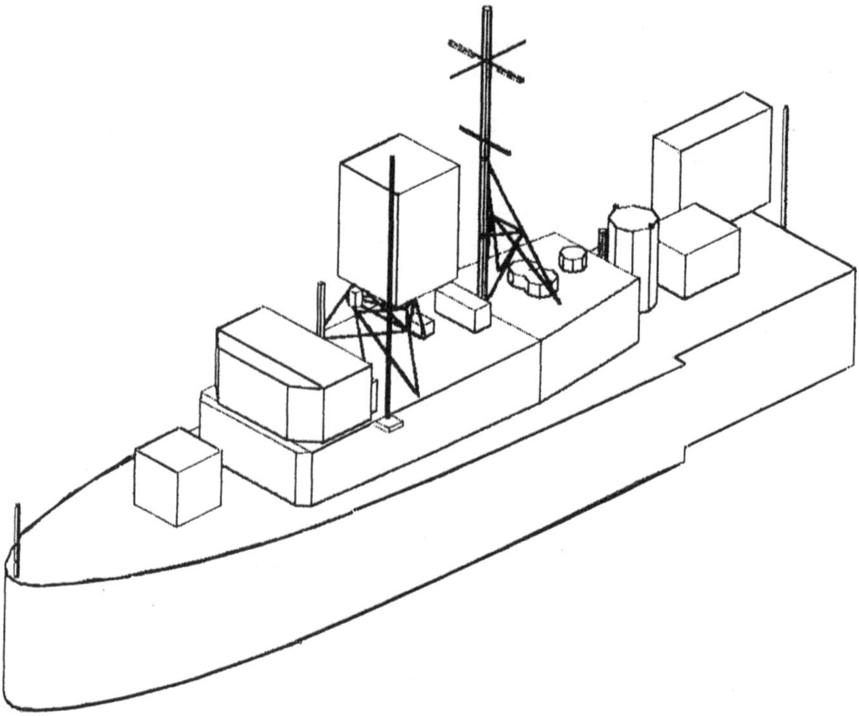

Figure 7-2 Isometric View of PHM 1 Class Ship TDM

7-2.3.2 TDM Omnidirectional Antenna Performance Evaluation

Shipboard omnidirectional antennas are used mainly for HF, VHF, and UHF communications and navigation. Examples include monopoles, wire-rope fans, discone cages, and vertical dipoles. EM performance is evaluated in terms of the desired radiation characteristics and predicted communication range. The antenna radiation pattern, degraded by the ship structure as shown in Figure 7-3, is one important measure of probable communication range at a specific bearing as viewed from the subject antenna, and is essentially determined by the geometry of the antenna itself, the geometry of the nearby superstructure acting as obstacles, reradiators, or reflectors, and the relative geometrical configuration of the composite antenna and superstructure taken as a whole.

To evaluate the radiation pattern qualitatively, a statistical or probabilistic approach is adopted. Namely, the radiation pattern is converted into a set of statistical descriptions; i.e., the amplitude probability distribution showing the antenna gain versus the number of degrees at which the level (gain) is exceeded

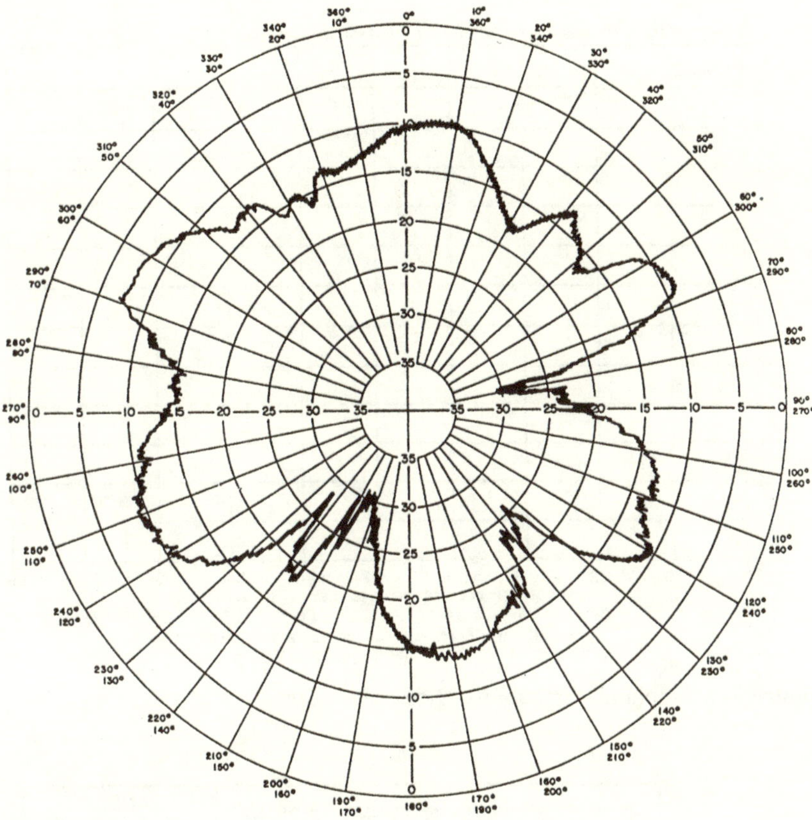

Figure 7-3 Communications Antenna Radiation Pattern Disturbed by Ship Structure

for a given probability. Figure 7-4 provides the amplitude probability distribution curve for a particular HF groundwave antenna. The curve is obtained as follows: For the azimuthal pattern (both vertical and horizontal polarization) at zero-degree elevation, the amplitude of the radiation level is sampled at each degree of azimuth. Then, 360 sampled values are arranged in descending order to form the amplitude distribution. Eleven equally spaced values taken from this distribution are selected and listed as shown in Figure 7-4.

In this statistical reduction, the number and azimuthal location of nulls and peaks are lost, but the cumulative amplitude probability distribution curves, thus obtained, enable one to compare quantitatively the radiation pattern of various omnidirectional antennas for given shipboard installations, as depicted in Figure 7-5.

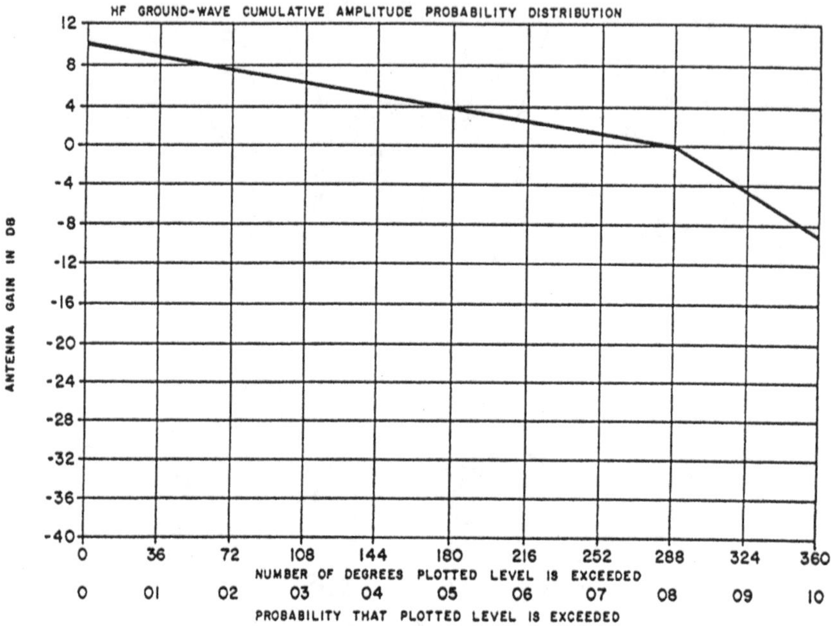

Figure 7-4 Amplitude Probability Distribution Curve

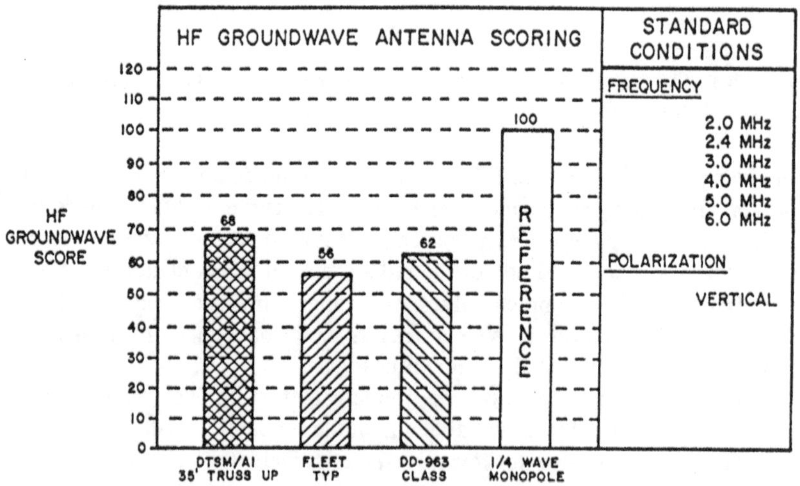

Figure 7-5 Antenna Scoring Comparison Summary

The omnidirectional antenna communication range is predicted in terms of the probability of successful communications of the HF, VHF, and UHF transmitting and receiving antennas.

HF surface-wave communication range is essentially dependent on the individual equipment parameters, antenna radiation pattern, atmospheric noise, and signal propagation characteristics. The equipment parameters include transmitted power, required receiver signal-to-noise ratio, and type of modulation. Again, the radiation pattern is dependent on the operating frequencies and the proximity of neighboring structures, and is described as a complementary cumulative amplitude distribution function. HF surface-wave communication range is limited primarily by atmospheric noise. Man-made noise, galactic noise, and receiver noise are considered secondary. For our illustration purposes, the HF communication antenna is assumed to be vertically polarized and located at a specified height above the design waterline. With these parameters, HF surface-wave communication range versus probability of successful communication is calculated.

For VHF and UHF line of sight, the successful communication range depends primarily on the receiver noise level, which is the limiting factor for circuit operation. Otherwise, the transmitter and receiver parameters are similar to those of the HF communication antenna. If the communication range is within the reflection region, the free space propagation loss is used as an approximation. Figure 7-6 provides VHF communication ranges versus probability of successful communication.

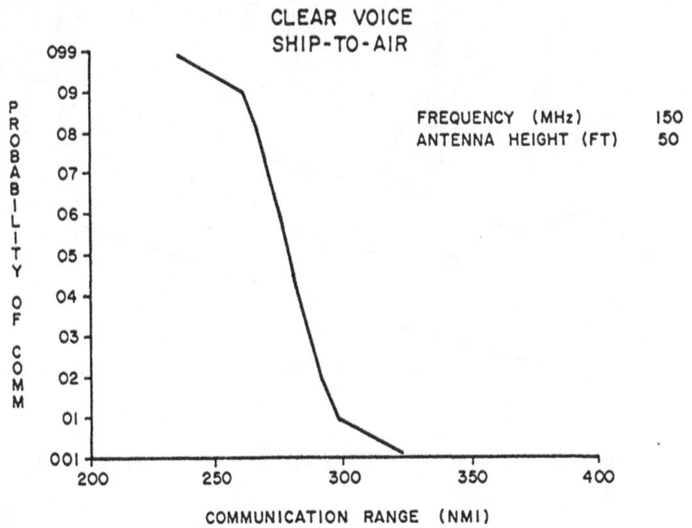

Figure 7-6 VHF Communication Range

7-2.3.3 TDM Directive Antenna Performance Evaluation

Directive shipboard antennas transmit and receive spatially concentrated electromagnetic energy (beams) in one direction at a time. Examples are air and surface search radars, gunfire control radar systems, and satellite communication systems. Their EM performances are evaluated in terms of optical coverage and radar detection range.

In locating a particular directive antenna in the shipboard topside, a knowledge of optical or ''geometrical'' coverage as viewed from the vantage point (the point specified at the antenna site) is essential. Especially at microwave frequencies, the optical coverage is considered sufficient as a first-order approximation to predict directive antenna EM performance. Figure 7-7 shows the isometric view of a hydrofoil candidate ship for new patrol boats. The optical view from a weapon system located at the midship is shown in Figure 7-8. The aft portion represents blockage caused by the weapon, while the forward shadow corresponds to blockage caused by the mast. A structural blockage caused by the pilot house appears forward while the vertical strip portions represent blockages caused by the HF whips.

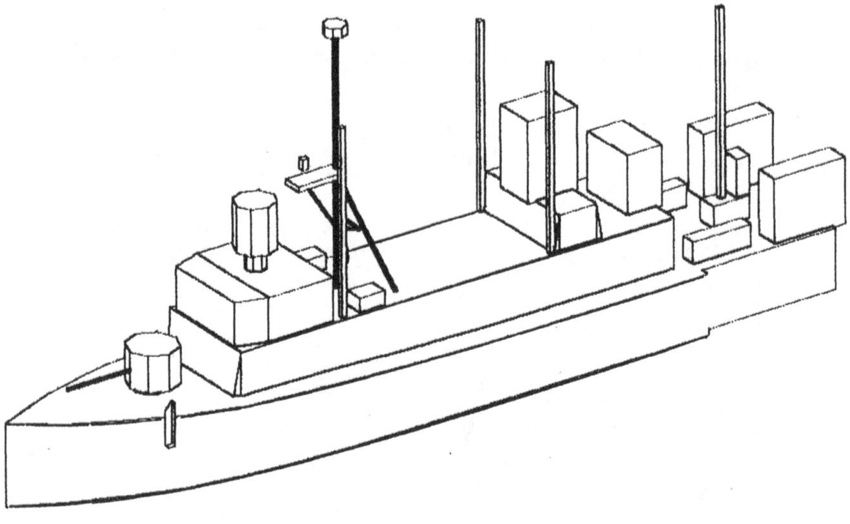

Figure 7-7 Isometric View of Hydrofoil Candidate Patrol Boat TDM

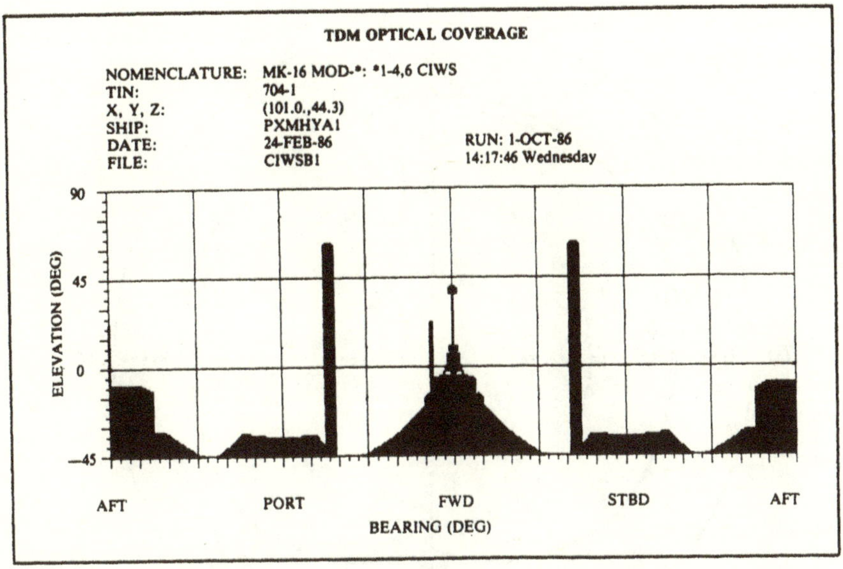

Figure 7-8 Weapon System Optical Coverage

Radar detection range is a function of the radar line-of-sight path and the cumulative radar system performance losses. These losses include optical blockage loss (gain reduction at a given azimuthal angle), EMI loss, jamming loss, and transmission line loss. Optical blockage loss is due to the geometry of the nearby superstructure and the relative configuration between the radar and superstructure. EMI loss, jamming loss, and transmission line loss contribute to the radar systems performance degradation and can be evaluated using the performance loss data.

Radar line-of-sight distance depends on the technical characteristics of the particular radar equipment and target height above the water surface. Because the atmospheric refractive index causes electromagnetic waves to travel more slowly near the earth's surface than at higher levels, the propagation speed variation results in a bending of the radar beam so that the radar is often able to detect the target beyond the horizon. To compensate for this beam bending effect, the earth's radius is increased by a factor of 1.5 in the radar line-of-sight distance calculation. When the system performance data and the radar line-of-sight distance data are entered into the TDM program, the radar detection range is displayed on a polar diagram, as depicted in Figure 7-9.

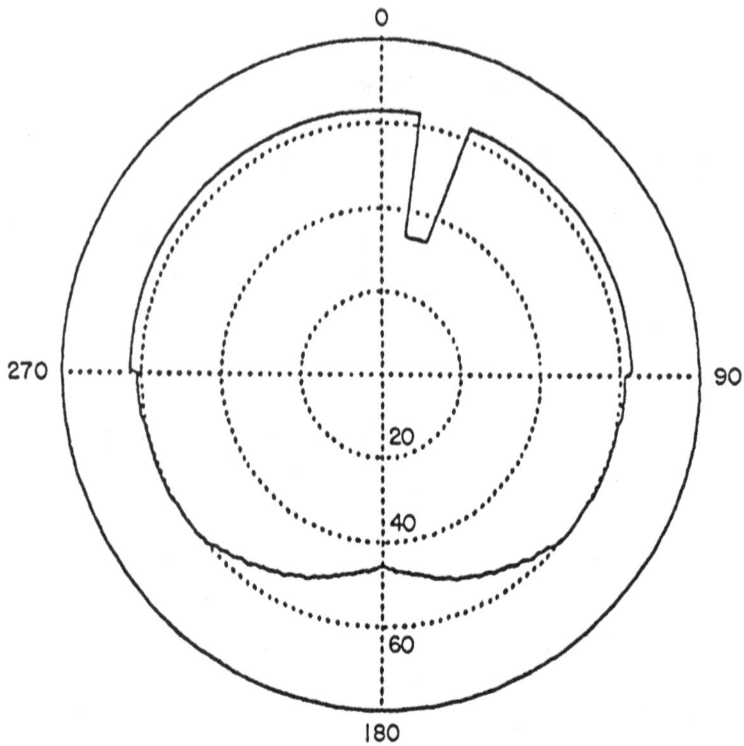

Figure 7-9 Radar Detection Range

7-2.3.4 Shipboard EM Assessment Summary

Integration of various topside elements into the complicated shipboard environment is a difficult engineering task, especially for warships with severely limited real estate. To accomplish this demanding task, TDM is being used effectively. For example, Figures 7-10 and 7-11 illustrate two recently modeled hulls produced by the TDM process for a modern naval destroyer and an amphibious landing ship. The predicted coverages for a major weapon system (controlled electromagnetically) are shown for the two hulls in Figures 7-12 and 7-13, respectively. In Figure 7-14 the overlapping (complementary) total coverage of three identical weapons on an aircraft carrier is depicted in polar view.

Figure 7-10 TDM Isometric Destroyer Hull

Figure 7-11 TDM Isometric Amphibious Landing Ship Hull

The central issue in shipboard electromagnetics is performance, specifically the ability to overcome performance degradation in the presence of interference. The topside design objective is to provide optimum overall performance as an integrated combat system in support of required ship missions. The overall combat performance must be effective in coverage, range, and reaction, yet be free of EMI, and not be an electromagnetic hazard to personnel, ordnance, or fuel. TDM technology has afforded improved flexibility and service to reach this aim.

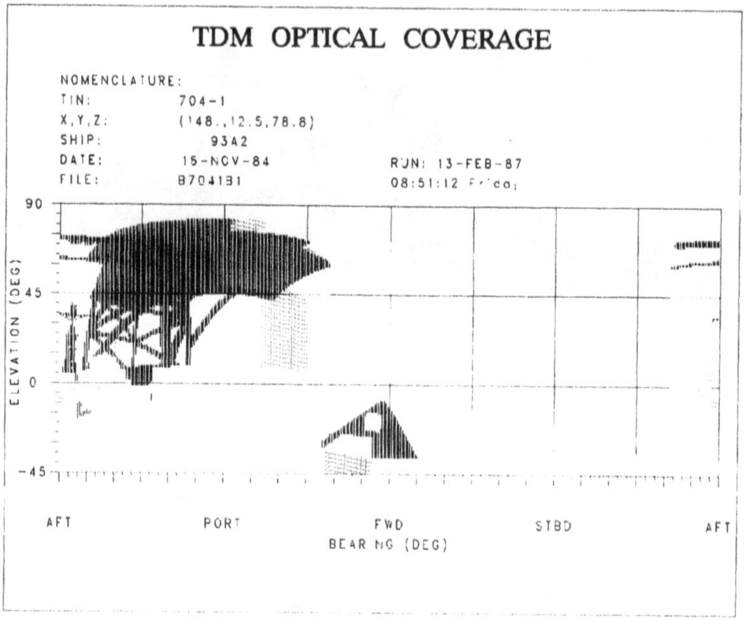

Figure 7-12 TDM Weapon Coverage (Ship Structure Blockage)

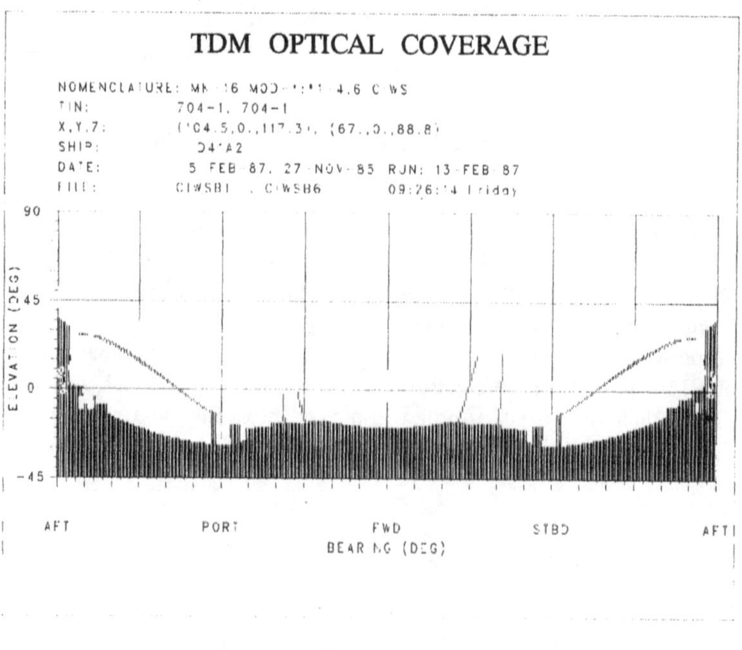

Figure 7-13 TDM Weapon Coverage (Ship Structure Blockage)

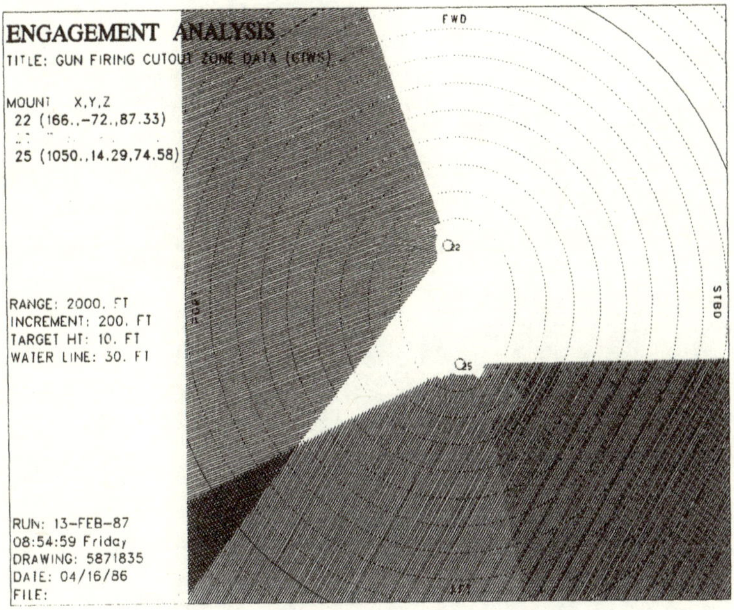

Figure 7-14 TDM Tri-Weapon Complementary Coverage

REFERENCES

1. B. E. Keiser, *Principles of Electromagnetic Compatibility*, Artech House, Dedham, MA, 1979, pp. 208-209.
2. F. M. Prout, R. C. Baker, and H. J. DeMattia, Jr., "Combatant Capability Assessment Status in the Ship Design Process," *Naval Engineers Journal*, June 1974, pp. 56-64.
3. Keiser, *op. cit.*, p. 212.
4. P. E. Law, Jr., S. Kuniyoshi, and T. P. Morgan, "Combatant Patrol Boat Topside Design," *Naval Engineers Journal*, January 1987, pp. 59-69.

Glossary

Acronym	*Definition*
ACGIH	American Conference of Governmental Industrial Hygenists
ASAP	Antenna Scattering Analysis Program
AVGAS	Aviation Gas
CCB	Change Control Board
CCS	Central Control Station
CONREP	Connected Replenishment
ECAC	Electromagnetic Compatibility Analysis Center
ECM	Electronic Countermeasures
ECP	Engineering Change Proposal
EED	Electroexplosive Device
EHF	Extremely High Frequency
EMC	Electromagnetic Compatibility
EMI	Electromagnetic Interference
EMCAB	Electromagnetic Compatibility Advisory Board
EMCON	Emission Control
EMCPP	Electromagnetic Compatibility Program Plan
EME	Electromagnetic Environment
EMP	Electromagnetic Pulse
EMPAL	Electromagnetic Pulse Design Algorithm
EMPRESS	Electromagnetic Pulse Radiation Environment Simulator for Ships
EMR	Electromagnetic Radiation
EPA	Environmental Protection Agency
EW	Electronic Warfare
HEMP	High-Altitude Electromagnetic Pulse

HERF	Hazards of Electromagnetic Radiation to Fuel
HERO	Hazards of Electromagnetic Radiation to Ordnance
HERP	Hazards of Electromagnetic Radiation to Personnel
HF	High Frequency
IF	Intermediate Frequency
IFF	Identification, Friend or Foe
MOGAS	Motor Vehicle Gasoline
NEMP	Nuclear Electromagnetic Pulse
OR	Operational Requirement
PEL	Permissible Exposure Levels
PEP	Performance Evaluation Program
RADHAZ	Radiation Hazards
RAM	Radar Absorbent Material
RF	Radio Frequency
RFI	Radio Frequency Interference
RLC	Resistance, Inductance, and Capacitance
SAR	Specific Absorption Rate
SATCOM	Satellite Communication
SEMCAC	Shipboard Electromagnetic Compatibility Analysis - Communications
SEMCAM	Shipboard Electromagnetic Compatibility Analysis - Microwave
SEMCIP	Shipboard Electromagnetic Compatibility Improvement Program
SHF	Superhigh Frequency
SHIPALT	Ship Alteration
TACAN	Tactical Air Navigation
TDIET	Topside Design Integration Engineering Team
TDM	Topside Design Model
TEMP	Test and Evaluation Master Plan
TESS	Tactical Electromagnetic Systems Study
TESSAC	Tactical Electromagnetic Systems Study Action Council
TPD	Terminal Protection Device
UHF	Ultrahigh Frequency
VERTREP	Vertical Replenishment
VHF	Very High Frequency
VSWR	Voltage Standing Wave Ratio
WCAP	Waterfront Corrective Action Program

BIBLIOGRAPHY

The following texts are a representative sampling for readers desiring further study in electromagnetics, from the beginning with Oliver Heaviside's majestic original treatment of modern-day engineering principles and applications to the present.

Ghose, R. H., *EMP Environment and System Hardness Design,* Don White Consultants, Gainesville, VA, 1984.

Hayt, W. H., *Engineering Electromagnetics,* McGraw-Hill Book Company, New York, 1967.

Heaviside, O., *Electromagnetic Theory,* Dover Publications, New York, 1950 (first published in London, 1899–1912).

Keiser, B. E., *Principles of Electromagnetic Compatibility,* Artech House, Norwood, MA, 1987.

Kraus, J. D., *Electromagnetics,* McGraw-Hill Book Company, New York, 1953.

Messenger, G. C., and M. S. Ash, *The Effect of Radiation on Electronic Systems,* Van Nostrand Reinhold Company, New York, 1986.

Ott, H. W., *Noise Reduction Techniques in Electronic Systems,* John Wiley & Sons, New York, 1976.

Panofsky, W. K. H., and M. Phillips, *Classical Electricity and Magnetism,* Addison-Wesley Publishing Company, Reading, MA, 1962.

Paris, D. T., and F. K. Hurd, *Basic Electromagnetic Theory,* McGraw-Hill Book Company, New York, 1969.

Ramo, S., J. R. Whinnery, and T. Van Duzer, *Fields and Waves in Communication Electronics,* John Wiley & Sons, New York, 1967.

Shelkunoff, S. A., *Electromagnetic Waves,* D. Van Nostrand Company, New York, 1943.

Index